MY REVISION NOTES

CCEA

GCSE

GEOGRAPHY

SECOND EDITION

Gillian Rea

Jennifer Proudfoot

Boost

HODDER
EDUCATION
AN HACHETTE UK COMPANY

The Publishers would like to thank the following for permission to reproduce copyright material.

Photo credits

p7 © Chris Radburn/PA Images/Alamy Stock Photo; **pp9–11** © Crown copyright and database rights 2021 Hodder Education under licence to OS; **p10** © Crown copyright (2021) permit number 180030; **p14l** © Geogphotos /Alamy Stock Photo, **p14r** © Kevin Britland/Alamy Stock Photo; **p19** © David Bagnall/Alamy Stock Photo; **p24** © Travel and Landscape UK/Mark Sykes/Alamy Stock Photo; **p30** © Mr. Nut/Alamy Stock Photo; **p31** © Ernie Janes/Alamy Stock Photo; **p36** © geogphotos/Alamy Stock Photo; **p.39** © Weerapat1003/stock.adobe.com; **p40l** © Eye Ubiquitous/Alamy Stock Photo; **p40r** © Crown Copyright, Met Office, Data: NASA; **p48t** Image: © Crown copyright, Met Office, Data: © NASA, **b** © NASA/Alamy Stock Photo; **p49b** Image from metoffice.gov.uk Contains public sector information licensed under the Open Government Licence v3.0, **t** Jacques Descloitres, MODIS Land Rapid Response Team, NASA/GSFC; **p75t** © Crown copyright and database rights 2021 Hodder Education under licence to OS, **m** © Neil Mitchell/Shutterstock, **bl, br** © Commission Air/Alamy Stock Photo; **p76** © Crown copyright and database rights 2021 Hodder Education under licence to OS, **p77** © Christopher Hill/scenicireland.com/Christopher Hill Photographic/Alamy Stock Photo; **p78** © John M Fisher/Shutterstock.com; **p79t** © John Takai/stock.adobe.com, **b** © John M Fisher/Shutterstock.com; **p81** © Alan R Gallery/Alamy Stock Photo; **p84** Matt-80 /Wikimedia Commons/Creative Commons Attribution 2.0 Generic license; **p91** © Hippo Roller/hipporoller.org; **p104** © leodaphne/Shutterstock; **p109t** © geogphotos/Alamy Stock Photo, **b** © Kumar Sriskandan/Alamy Stock Photo; **p111** © geogphotos/Alamy Stock Photo; **inside cover** © Crown copyright (2021) permit number 180030.

Maps and symbols included by kind permission of Ordnance Survey (OS). Ordnance Survey (OS) is the national mapping agency for Great Britain, and a world-leading geospatial data and technology organisation. As a reliable partner to government, business and citizens across Britain and the world, OS helps its customers in virtually all sectors improve quality of life.

Acknowledgements

p.8 NOAA OPC, Weather maps. Retrieved from https://ocean.weather.gov/Pac_tab.php; **p.80** Real Journey Time, Real City Size, and the disappearing productivity puzzle. © Tom Forth, The Productivity Insights Network.

Every effort has been made to trace all copyright holders, but if any have been inadvertently overlooked, the Publishers will be pleased to make the necessary arrangements at the first opportunity.

Although every effort has been made to ensure that website addresses are correct at time of going to press, Hodder Education cannot be held responsible for the content of any website mentioned in this book. It is sometimes possible to find a relocated web page by typing in the address of the home page for a website in the URL window of your browser.

Hachette UK's policy is to use papers that are natural, renewable and recyclable products and made from wood grown in well-managed forests and other controlled sources. The logging and manufacturing processes are expected to conform to the environmental regulations of the country of origin.

Orders: please contact Hachette UK Distribution, Hely Hutchinson Centre, Milton Road, Didcot, Oxfordshire, OX11 7HH. Telephone: +44 (0)1235 827827. Email education@hachette.co.uk Lines are open from 9 a.m. to 5 p.m., Monday to Friday. You can also order through our website: www.hoddereducation.co.uk

ISBN: 978 1 3983 2117 5

© Gillian Rea and Jennifer Proudfoot 2021

First published in 2018
Second edition published by
Hodder Education,
An Hachette UK Company
Carmelite House
50 Victoria Embankment
London EC4Y 0DZ

www.hoddereducation.co.uk

Impression number   10 9 8 7 6 5 4

Year   2025 2024

Cover photo © Martina – stock.adobe.com

Illustrations by Aptara, Inc.

Typeset in India by Aptara, Inc.

Printed in India

A catalogue record for this title is available from the British Library.

Get the most from this book

Everyone has to decide his or her own revision strategy, but it is essential to review your work, learn it and test your understanding. These Revision Notes will help you to do that in a planned way, topic by topic. Use this book as the cornerstone of your revision and don't hesitate to write in it – personalise your notes and check your progress by ticking off each section as you revise.

Tick to track your progress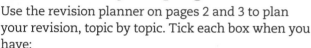

Use the revision planner on pages 2 and 3 to plan your revision, topic by topic. Tick each box when you have:
- revised and understood a topic
- tested yourself
- practised the exam questions and gone online to check your answers.

You can also keep track of your revision by ticking off each topic heading in the book. You may find it helpful to add your own notes as you work through each topic.

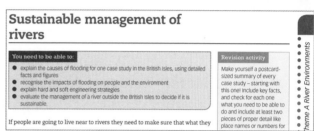

Features to help you succeed

Exam tips

Expert tips are given throughout the book to help you polish your exam technique in order to maximise your chances in the exam.

Now test yourself

These short, knowledge-based questions provide the first step in testing your learning.

Definitions and key words

Essential key terms are highlighted in bold throughout the book. Clear, concise definitions of these terms are provided in the Glossary on pages 120–25.

Revision activities

These activities will help you to understand each topic in an interactive way.

Exam practice

Practice exam questions are provided for each topic. Use them to consolidate your revision and practise your exam skills.

Online

Go online to check your answers to the exam questions and try the quick quizzes at **www.hoddereducation.co.uk/myrevisionnotesdownloads**

My revision planner

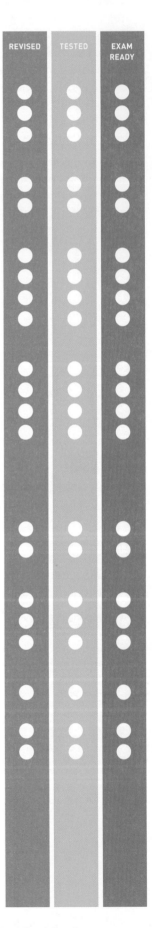

REVISED   TESTED   EXAM READY

Exam practice answers and quick quizzes at **www.hoddereducation.co.uk/myrevisionnotesdownloads**

My revision planner

Introduction

Why use this book?

Exam success depends on two things:
- your knowledge and understanding of the subject matter
- your ability to use that knowledge in the manner that will gain most marks in the examination.

To help you to gain your best possible grade, this book aims to:
- set out the subject content essential to the CCEA GCSE Geography course
- give you hints and revision tips that will help you to understand and memorise the material
- advise you on the best way to approach various types of exam questions.

Each chapter covers one of the eight themes making up the GCSE course, as well as providing now test yourself questions and exam questions so that you can check, as you go along, how much you understand and can remember. The Glossary on pages 120–25, provides clear and concise definitions of the key ideas required for each topic. This section is important as you will almost certainly be asked to define some of them in the examinations.

Examination techniques

Command words

To answer exam questions correctly, it is important to be sure what the examiner is really asking. Read the question carefully and underline the command words – these are words such as *state*, *describe* or *explain*. They tell you what to do in your answer. If you explain when asked to describe you will earn no marks, even if what you write is otherwise correct.

The following table gives the meanings of some of the command words you will meet.

Command word	Meaning
State	A short answer, presenting a fact or facts (for example, the temperature in January, taken from a graph)
Describe	A descriptive answer *without* trying to explain When describing a *graph*, it is important to *quote figures* When describing a *map*, it is important to mention *place names*
Explain	Give a reason or reasons
Describe and explain	Make descriptive statements and give the reasons why (for example, describe the pattern of rainfall shown on a map and explain why it falls there)
Label	Add labels to a diagram
Complete	Add information to a graph or a table so that it is complete. If you are completing a graph, make sure you follow the shading that has already been used. If a bar is coloured in completely, make sure you colour it completely rather than scribbling, if you want full marks
Match	Match statements that have been presented in the form of 'heads and tails'
State the meaning	Usually used for definitions. You need to show that you know what the term means
Suggest	This is used when there may be more than one correct answer and any relevant answer is acceptable

Exam practice answers and quick quizzes at **www.hoddereducation.co.uk/myrevisionnotesdownloads**

Structure of the examination

You will have three exams for your Geography GCSE.

Unit 1   40%

Understanding our Natural World

You must answer four questions, one on each theme:
- River Environments
- Coastal Environments
- Our Changing Weather and Climate
- The Restless Earth

Each question will be made up of several parts, including short and long answer questions.

Unit 2   40%

Living in our World

You must answer four questions, one on each theme:
- Population and Migration
- Changing Urban Areas
- Contrasts in World Development
- Managing our Environment

Each question will be made up of several parts, including short and long answer questions.

Skills for Unit 1 and 2

You might be asked to:
- Give a definition for a term.
- Explain a geographical feature or process.
- Analyse a graph or table of data, and describe what it shows, quoting data to help.
- Give reasons for the patterns in a graph.
- Analyse a photograph, map or diagram.
- Use an Ordnance Survey map to answer questions: using grid references, measuring distances, using a key, giving compass directions, or recognising specific geographical features that you have learnt about.
- Write about a case study, giving precise information such as place names and numbers, to answer a question.

Unit 3   20%

Fieldwork

You will need to take into the exam a summary of your fieldwork, and a table of your data, which you will hand in with your answers. You will be asked questions about each stage in the Enquiry Process, starting with the planning stages and working right through to the Evaluation.

Some of these questions will need short answers of just a few words. Others will require more extended answers. You will have to draw a graph of some of your data.

Skills for Unit 3

You need to answer questions very precisely, talking about your fieldwork specifically. This might include place names and detailed descriptions, such as how exactly you took measurements.

You need to select an appropriate type of graph, draw it accurately and precisely, using a ruler and a sharp pencil, with accurate labels for your axes.

Revision techniques

Here are some useful tips about revision:

- Ideally, revision should be on-going throughout the course. Don't leave it all to the days just before the exam.
- Case studies, in particular, should be memorised as you go along so that facts about each one are clear in your mind before you study the next.
- Revision is *not* just rereading your notes or the textbook.
- Revision should involve reworking the subject matter, perhaps into a spider diagram or by summarising into brief bullet points.
- Next, you have to memorise the material by repeating it to yourself, explaining it to someone else, writing a list or making a poster.
- Test yourself, to see how much your memory has retained, by writing out the bullet points or list, redrawing the diagram or explaining it all to someone – this time without the help of your notes or book.
- Visual forms of revision, such as spider diagrams and learning maps, can be a big help. They let you picture the key points, arrange them under headings and see connections.
- Case studies need careful revision, with facts, figures and places committed to memory. Writing the title of each one with bullet points covering the main ideas (including the facts, figures and place names) on a card will help you to concentrate on this aspect of the course. Gradually, you will build up a collection of cards, ready for last-minute revision on the eve of the exam.

Do	**Rework**	**Memorise**	**Write an answer**
Stage 1	*Stage 2*	*Stage 3*	*Stage 4*
• Do classwork • Do homework • Make notes highlighting key words and relevant facts: places, figures and names	• Change notes into spider diagrams or learning maps • Add own ideas/theory and make links with previous notes • Discuss your work with the teacher or a friend • Make a summary box and structure the information under meaningful headings • Find new ways of thinking about something, such as using thinking skills	• Commit to memory by repetition • Say the information out loud or make up a rhyme or tune • Explain your topic to a friend • Put the information up on posters around your room and move around when learning • Take short breaks • Close your notes • Recall information by writing it out	• Select the information which is relevant to answer a GCSE question • Compose an answer to the question

Exam practice answers and quick quizzes at **www.hoddereducation.co.uk/myrevisionnotesdownloads**

Exam questions

Recall questions

These are designed to test your knowledge. For example, 'State the meaning of the term earthquake'.

Data response questions

These questions provide you with visual clues, so make the best use of them.

Describing tables and graphs
Give the overall pattern, with figures to back it up, and any exceptions.

OS maps
Each year there will be an Ordnance Survey map in either Unit 1 or Unit 2. You will need several skills to answer the questions – you can see more detail on this on pages 9–11.

Photographs
You might need to:
- match up a photograph with a location on an OS map
- comment on or label a feature such as a land use zone, evidence of regeneration or a waterfall.

Hard rock

Plunge pool

Soft rock has been eroded more easily, undercutting the hard rock, leaving it unsupported

Weather maps

Check if there is a date: winter or summer may make a difference.

Look at the pattern of the isobars:

- High pressure in the centre – anticyclone, clear skies.
- Low pressure in the centre – depression, rain and wind.

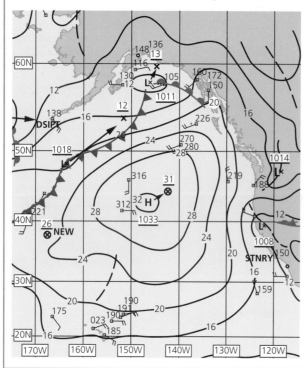

Case studies

For some level of response questions, you need to write about a specific place. For top marks, you should include at least two facts (figures, names or places).

Rather than writing: 'Lots of people were killed', you could say, '5500 people died'.

Rather than writing: 'Jobs were lost because factories closed', you might say, 'There was unemployment when the Mitsubishi factory closed'.

Levels of response questions

- Worth 6 marks or more.
- Require extended written answers.

For example: 'Explain two causes of flooding.'

To get full marks, for each of your causes you need to:

Make a statement

Building on the floodplain can cause flooding.

Give a reason or consequence

This is because it creates impermeable surfaces, so water cannot soak into the soil, but instead flows quickly over the surface to the river.

Elaborate on the reason or consequence

This can be further detail or an actual named example.

For example, in the Somerset Levels new houses were built in Taunton and Bridgewater, so water reached the River Parrett quickly and made it more likely to flood.

Exam practice answers and quick quizzes at **www.hoddereducation.co.uk/myrevisionnotesdownloads**

Ordnance Survey map questions

You will almost certainly get an Ordnance Survey map as part of your Unit 1 or Unit 2 exam, with several questions to answer.

You will be given a key with the map, showing you what the symbols mean. You should try to become familiar with these as much as possible, as it will make it much easier in the exam. You can see a full key at www. ordnancesurvey.co.uk/documents/50k-raster-legend.pdf

You need to be able to:

1   **Measure distances accurately** between two places:
● measure carefully with a ruler in cm (remember your ruler)
● always measure from the centre of a symbol
● convert your measurement into kilometres by dividing by 2.

2   **Use compass directions**: for example, farm A is north east of village B. You will only need eight compass directions – if a direction is in between them, pick the closest option. Be careful to make sure you go in the right direction, not the reverse. Put your finger on the starting point and work out in what direction you have to move it to get to your destination. Make sure you use a symbol if there is one, rather than a name.

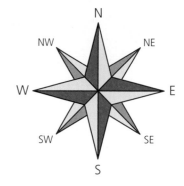

For example, in the map on the inside back cover, what direction is Cross Slieve from Cushendall?

Your starting point is Cushendall, so put your finger on the village. Find Cross Slieve, and move your finger towards it – you are travelling north. So Cross Slieve is north of Cushendall.

3   **Find out the height above sea level** from contour lines, spot heights or triangulation pillars and understand the contours showing hills and valleys.

● 31

| A **spot height** is a black number – sometimes with a tiny black dot beside it – to show the height. | **Triangulation pillar** – this will have a number next to it showing the height of the land. |

Contours are brown lines which join up the places that are the same height above sea level.

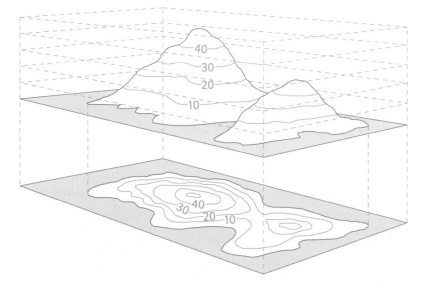

The numbers show the height of the land. Contours are drawn every 10m, so you can work out the height if you need to, by counting the lines from one that is numbered. In the example, you can count the contours from 600, so there is a contour at 610, 620, 630 and 640, then you reach the labelled contour at 650.

Patterns in the contours show the shape of the land.

Pattern	Shape of the land
No contours	Flat land
Contours close together	Steep slope
Contours far apart	Gentle slope
Contours in circular shape	Hilltop
V-shape, with lower numbers in the middle of the V	Valley

4 **Find features on the map** using four-figure and six-figure grid references.

All Ordnance Survey maps are divided up into grid squares, marked with blue lines. Grid references are like coordinates. A four-figure reference tells us which square to look at. The first two figures indicate the vertical line on the left of the square. The last two indicate the horizontal line on the bottom of the square.

If you have a six-figure grid reference, start by using the first two figures and the fourth and fifth figures to identify the correct square. The third figure shows how many tenths of the way across the square to go. The last figure shows how many tenths of the way up the square to go.

In an exam, you might be given a six-figure grid reference to help you find a specific feature.

> **Exam tip**
>
> To be really accurate, use a ruler to measure across and up the square. Each tenth is 2 mm.

5 **Identify features** from Unit 1 and Unit 2.

Unit 1: Identify river features and land uses, and coastal features and land uses.

Unit 2: Identify land use zones in a city.

> **Now test yourself** TESTED ◯
>
> Using the map on this page (a large version can be found on the inside back cover of this book):
>
> 1 What shape is the land next to the river in grid square 2331?
>
> 2 What height is the land at 228276?
>
> 3 Using map evidence, suggest one use of the coast in grid square 2427.
>
> 4 What direction is Cushendall (2427) from Gruig Top (2030)?
>
> 5 What river features can you identify in grid square 2332?
>
> 6 Using the map on page 75, **Figure 2** Ordnance Survey map of Sheffield: Measure the distance from the bus station at 358872 to the nature reserve at 315852
>
> 7 Identify the land use zone in grid square 3085.
>
> 8 Give one piece of evidence that shows square 3587 is part of the CBD.

> **Exam tip**
>
> Use the centre of each symbol.

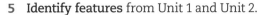

Four-figure grid references

Square A is 2180
Square B is 2279

Six-figure grid references

Point P is 21**5**80**5**
Point Q is 22**8**80**2**
Point R is 22**0**79**5**
Point S is 22**5**79**7**

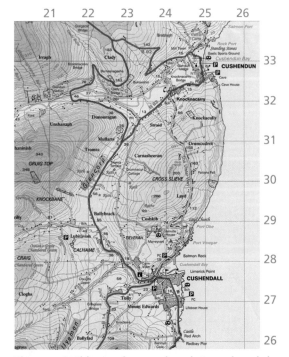

Please note this map is reproduced at a reduced size. The map on the inside back cover is shown at 100%.

Unit	Theme	Feature to identify	What to look for
Unit 1	River environments	Waterfall	River going across lots of contours together, usually labelled 'Falls' or 'Waterfalls'
		Floodplain	No contours either side of a river showing flat land
		Meander	Shape of the river, showing bends
		Sewage works	Usually marked with circles, often labelled 'works'
		Factories	May be labelled 'works'
		Roads and railways along the valleys	Red/yellow lines, black railway lines following along the river
	Coastal environments	Headland	Look for the shape – land sticking out to sea. Often has the word 'head' in its name, like Ramore Head in Portrush
		Cliff	Look for vertical cliff symbols (see below)
		Wave-cut platform	Look for rocks marked in the water (see below)
		Cave, arch	These are often not marked, but if there are lots of them they may be labelled
		Stack, stump	These may appear as tiny islands just next to the coast
		Sandy beach, shingle beach	Look for yellow (sand) and yellow or white with speckles (shingle) along a coast (see below)
		Spit, hooked spit	Look for the shape – long and thin, stretching out from the land, often with sand or shingle marked along it
		Port or harbour	Look for a wall shape going out into the sea, with lighthouses or ferry routes
		Tourism	Look for blue tourism symbols, parking, nature reserves, golf courses

For Unit 2 features, see page 76.

The drainage basin: a component of the water cycle

> **You need to be able to:**
> - understand how water moves around in the drainage basin
> - identify and define parts of a drainage basin
> - explain what changes occur along the long profile of a river, and why.

Components of the drainage basin

REVISED ⬤

The water cycle is the way water is evaporated from the sea, goes through the air and flows back to the sea through rivers or the ground.

A drainage basin is the area of land drained by a river and its tributaries. In other words, any rain that falls in a particular area of land will end up in one particular river.

The rain may have a very eventful journey before it reaches the river. You need to understand the different parts of that journey. We sometimes talk about the parts of the journey as a system – just like the way food goes through our digestive system.

All systems have:
- things which go into them (inputs)
- ways of moving something from one place to another (transfers)
- places where things are stored (stores)
- things which come out at the end (outputs).

Revision activity

Make yourself a set of flashcards using the table below with the names on one side and the definitions on the other. Colour-code them to represent inputs, transfers, stores and outputs. Get someone to test you on them. Then use them to create a giant diagram like Figure 1.

	Name	Meaning
Input	**Precipitation**	Any water falling from the sky: rain, snow, sleet, hail
Stores	**Interception** by vegetation	Leaves and grass catch raindrops as they fall, and store them. Try sheltering under a tree next time you get caught in the rain! But don't stay too long – if there's too much water stored on a leaf it can fall to the ground
Transfers	**Surface runoff/overland flow**	Water running over the surface of the ground
	Infiltration	Water sinking into the soil
	Throughflow	Water flowing through the soil
	Percolation	Water sinking down through the rock
	Groundwater flow	Water flowing slowly from the rock into the river
Outputs	River discharge	Water flowing away in the river
	Evapotranspiration	Water turning into water vapour in the air, and water turned into water vapour by plants through their leaves

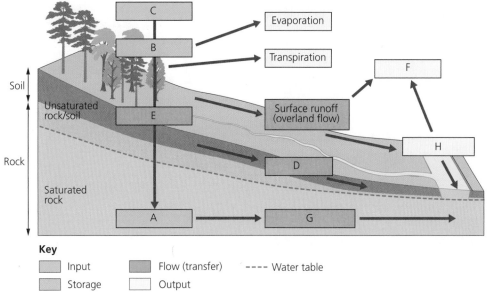

Key

▓ Input	▓ Flow (transfer)	---- Water table
▓ Storage	▢ Output	

Figure 1 The drainage basin system.

Now test yourself   TESTED ○

1 Match the words from the middle column of the table on page 12 to the boxes A–H on Figure 1.
2 Name one input into the drainage basin cycle.
3 What term means that trees and other plants catch raindrops as they fall?
4 What term means water flowing slowly from rock into the river?
5 What is meant by the term 'percolation'?
6 Is river discharge an input, transfer or output of the drainage basin cycle?
7 What term means water returns to the air from the drainage basin as vapour?
8 What is the difference between infiltration and throughflow?

Features of a drainage basin   REVISED ○

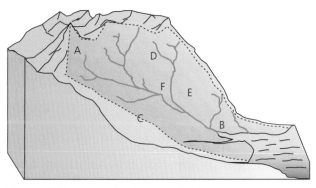

Figure 2 Features of a drainage basin.

Exam tip

In the exam, you could be given a diagram to label with these features. You also need to use technical terms like these when you are explaining things.

Now test yourself   TESTED ○

Match the letters A–F on Figure 2 to the following labels:

1 **Drainage basin**: the area of land drained by a river and its tributaries.
2 **Source**: the place where a river starts.
3 **Tributary**: a stream flowing into a river.
4 **River mouth**: where a river flows into the sea.
5 **Watershed**: the boundary between drainage basins – often a ridge of high land
6 **Confluence**: where two streams or rivers meet.

13

Changes along the long profile of a river

The long profile of a river means its shape from the source to the mouth. Imagine cutting down through the land to be able to see the whole river from source to mouth.

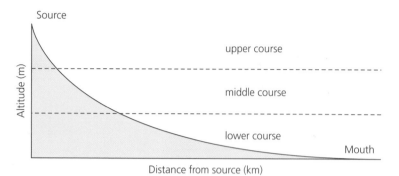

Figure 3 Sketch of long profile of river.

	Gradient	Depth	Width	Discharge	Load
Meaning	The steepness of the slope the river flows down	Measure from top of water to river bed. Take average across river	Distance from one side of the river to the other	Amount of water passing a point in a certain time – cumecs (cubic metres of water per second)	The material a river is carrying – mud, sand, pebbles, rocks
Change as you go downstream	Gets less steep	Gets deeper	Gets wider	Increases	Particles get smaller and more rounded
Why?	The river does more downwards erosion near the source, and more sideways erosion near the mouth	The river erodes downwards as it travels (vertical erosion)	The river erodes sideways as it travels (lateral erosion)	More water flows into the river from each tributary Water flows faster with less friction	Particles knock against each other and break each other up. Sharp angular edges get knocked off

Figure 4 River channel changes along the long profile.

Figure 5 Source and mouth of a river.

Exam practice answers and quick quizzes at **www.hoddereducation.co.uk/myrevisionnotesdownloads**

TESTED ◯

1 Complete the table below using the following words: Angular, Deep, Gentle, High, Large, Low, Narrow, Rounded, Shallow, Small, Steep, Wide.

	Near source	Near mouth
Gradient (how steep the land is)		
Width (from bank to bank)		
Depth (from water surface to the river bed)		
Discharge (amount of water going past a certain point in a second)		
Load (particles carried by the river) (two words for each column)		

2 What is the term used for a stream that flows into a river?

3 What is the term used for the point where two streams or rivers meet?

4 What does the term 'watershed' mean?

Revision activity

Try drawing this as a Bradshaw model diagram. For each characteristic of the river, draw a horizontal line, getting wider or narrower as it goes across the page, to represent the change.

For example:

Width

Exam practice

1 Study Table 1, which shows how the Whitewater River in the Mourne Mountains changes downstream. Answer the questions that follow.

Table 1

Distance from source (km)	Width of river channel (m)	Depth of river channel (m)	Size of load, longest axis (cm)
1	1.27	0.05	14
17	12.20	0.24	7

a) Describe how the river channel changes downstream. [4]

b) The load is smallest near the mouth of the river. State fully **one** reason why this is so. [3]

River processes and landforms

> **You need to be able to:**
>
> ● understand erosion, transportation and deposition
> ● understand how waterfalls, meanders, floodplains and levees are formed, using diagrams
> ● use aerial photographs and OS maps to identify river landforms and land uses.

Processes of erosion, transportation and deposition

REVISED

Erosion – breaking up and removing land

Deposition – dropping the load

Transportation – carrying along eroded material (load)

Figure 6 River processes.

A more detailed explanation of the processes of erosion can be seen in Figure 7 – the same processes work at the coast and in rivers.

> **Now test yourself**
>
> TESTED
>
> 1 For each of the following descriptions, decide what process is happening:
>   a) Large pieces of rock are broken off the edge of the river bank.
>   b) The river looks very brown because it is carrying lots of mud.
>   c) During a flood the river gets wider.
>   d) There is a tiny beach at the edge of the river where it flows into a lake.
>   e) If you paddle barefoot in the river you may get stones hitting against your feet.
>   f) Pebbles on the river bed are sharp near the source and smooth near the mouth.
> 2 What happens to large angular rocks and pebbles as a river carries them downstream?
> 3 What erosion process is the action of water plus the load it is carrying?
> 4 What river transport process makes stones bounce or hop along the river bed?
> 5 What happens to a river's load when the river slows down?

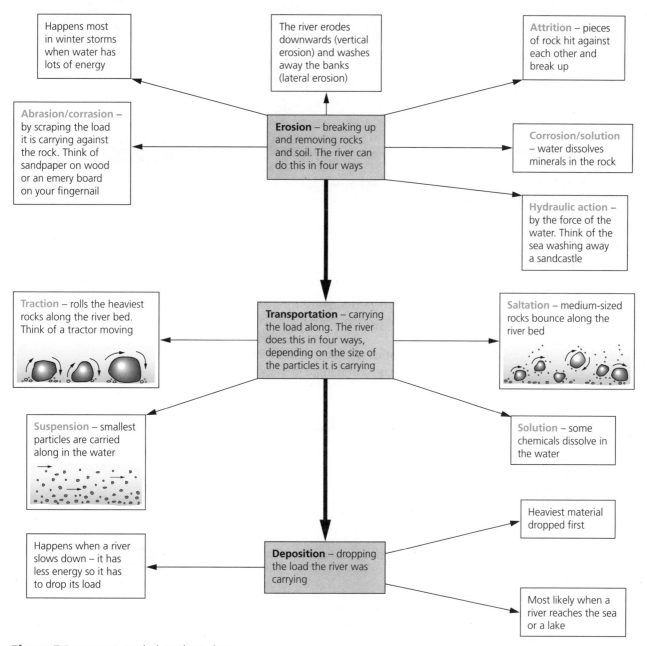

Figure 7 Processes carried out by a river.

Revision activity

Do an internet search to find an animation called 'Sticky does River Processes'. This is a brilliant clip using cartoons to explain all the processes you need to know. Watch the clip, then try to draw your own cartoons to illustrate all the words in bold in Figure 7. The funnier you make them, the more likely you are to remember what they mean!

Formation of river landforms

You need to be able to use annotated diagrams to explain how waterfalls, meanders, floodplains and levees are formed, and identify these river landforms on maps and aerial photographs.

Formation of a waterfall

- A waterfall is formed when there is a layer of **hard rock** on top of a layer of **soft rock**.
- The river erodes the **soft rock** more easily, so there is a step in the river bed. Eventually this becomes deeper, making a waterfall.
- **Hydraulic action and abrasion make a plunge pool** at the bottom of the waterfall.
- More erosion **undercuts** (or cuts under) the hard rock, leaving it hanging over the plunge pool.
- The **overhanging hard rock falls into the plunge pool**, and the position of the **waterfall moves backwards**.

On an Ordnance Survey (OS) map, most waterfalls are marked with a written label saying 'Falls' or 'Waterfall'. Have a look at the OS map on the inside of the back cover of this book, grid square 1931 and 2024.

> **Exam tip**
>
> Explain each stage of the formation of a river landform carefully in an exam answer, including the names of the erosion processes, using all the technical terms in bold.

> **Now test yourself**   TESTED ○
>
> Match the words in **bold** above with the numbers in Figure 8 to label the diagram.
>
>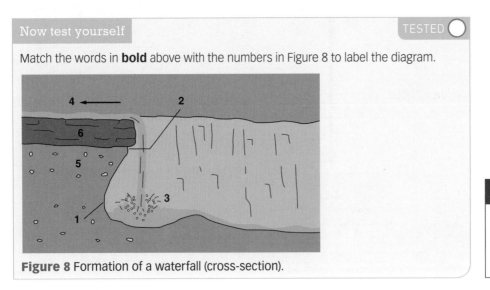
>
> **Figure 8** Formation of a waterfall (cross-section).

> **Exam tip**
>
> In an exam, you might have to label any of these features on a diagram or a photo.

Formation of a meander

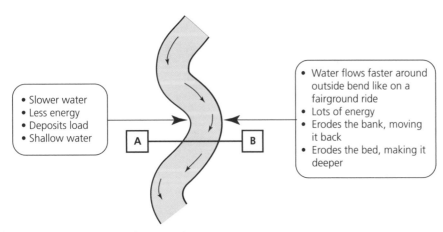

- Slower water
- Less energy
- Deposits load
- Shallow water

A

B

- Water flows faster around outside bend like on a fairground ride
- Lots of energy
- Erodes the bank, moving it back
- Erodes the bed, making it deeper

Figure 9 Formation of a meander.

Figure 10 shows what it would look like if you could cut through the river along the line marked A–B (a cross-section).

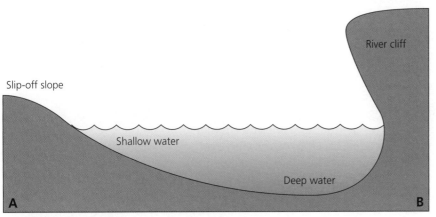

Figure 10 Cross-section through channel along the line A–B.

Now test yourself

TESTED

1 Describe the formation of a meander.
2 Draw a cross-section diagram of C–D from Figure 11. Label:
   - fast-flowing water
   - deep water
   - erosion
   - **river cliff**
   - slow-flowing water
   - shallow water
   - deposition
   - **slip-off slope**.

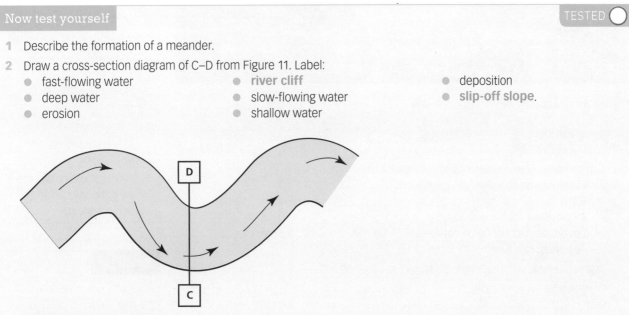

Figure 11 River meander.

3 Look at the OS map on the inside back cover of this book. In the north of the map, west of Cushendun, you will find a river with several meanders. Draw the shape of the river, and label the places where (a) erosion and (b) deposition will take place.

4 Study the photo below of the River Severn. You should be able to spot areas where deposition has happened. What colour do you see there?

Formation of a floodplain and levees

Flat land either side of a river will be covered in water if the river bursts its banks: this is called a floodplain.

Where there are meanders, the river flattens the land by eroding it and depositing sediment on it, making a flat floodplain. If the river gets too full and bursts its banks, water floods over the floodplain.

The water slows down, loses energy and deposits load (called sediment or alluvium), which is fertile (good for growing crops). The heavier, larger particles are deposited first, building up on the riverbank, while the smaller particles can be carried further across the floodplain. This means that over time the riverbanks will build up into long mounds, called levees, made of the larger particles.

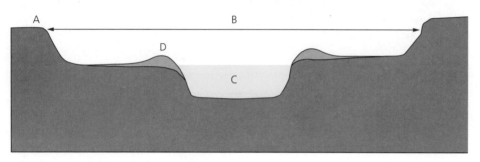

Figure 12 Cross-section of a floodplain.

Now test yourself

TESTED ◯

1 Use the definitions below to help you label letters A–D in Figure 12.
   - **River channel**: where the river flows.
   - **Bluff**: remains of higher land, where the rest has been eroded.
   - **Floodplain**: flat area covered in water in a flood.
   - **Levee**: banks built up on either side of the river where larger particles are deposited in a flood.
2 Name the feature found at the bottom of a waterfall.
3 True or false? A river erodes on the outside bank of a meander bend because the water is deep and travelling fastest there.
4 What is deposited on a floodplain as a result of flooding?

Exam practice

1 State the meaning of the term 'watershed'. [2]
2 Explain how a waterfall is formed. [5]
3 Using the map on the inside back cover of this book, identify two landforms which can be seen in grid square 2332. [2]
4 Using the map on the inside back cover of this book, explain why there is so little human land use on the floodplain in grid square 2323. [2]

Exam tip

For question 2, make sure you explain how a waterfall is formed step by step, using technical terms wherever you can.

Exam tip

In an exam you might have to label (annotate) a diagram or photo with the correct technical terms.

Revision activity

Look at the OS map on the inside back cover of this book. The floodplain of the Glenariff River stretches to the south-west of Glenariff or Waterfoot, grid square 2425.

1 What colour is the land in the floodplain shaded on the map?
2 What height is the land? (Look at the contour lines.)
3 How can you tell it is flat?
4 What do you think are the land uses on this floodplain?
5 What problems will there be if the river floods on its floodplain?

Exam tip

A legend and a scale will be provided with all OS maps that appear in the exam. You may be asked to identify features on maps or measure distances.

Revision activity

Practise drawing diagrams with labels for each landform.

Sustainable management of rivers

You need to be able to:

- explain the causes of flooding for one case study in the British Isles, using detailed facts and figures
- recognise the impacts of flooding on people and the environment
- explain hard and soft engineering strategies
- evaluate the management of a river outside the British Isles to decide if it is sustainable.

Revision activity

Make yourself a postcard-sized summary of every case study – starting with this one! Include key facts, and check for each one what you need to be able to do and include at least two pieces of proper detail like place names or numbers for each of these.

If people are going to live near to rivers they need to make sure that what they do to that river is sustainable. For example, if building a dam causes problems further down the river, then it is not a sustainable form of management.

Physical and human causes of flooding

REVISED ○

Flooding occurs when the water in a river channel is higher than the riverbank, so it overflows. Most rivers flood regularly and this may result from physical causes (such as heavy rainfall) or from human activity (such as building on the floodplain).

Case study: Causes of flooding – River Don, South Yorkshire, 2019

Physical causes	Human causes
• The area received 1.7 times the average rainfall from September to November, meaning that the soil was saturated, so any further rain would flow rapidly over the surface to the river.	• Intensive animal grazing upstream results in short grass and compacted soil, which limits the amount of water that can sink into the soil for storage. This means more of the rainfall flowed rapidly over the surface to the river.
• Sheffield had 84 mm of rain in the 36 hours that preceded the flood, almost a month's rainfall.	• Flood defence funding was cut by 30% in Yorkshire in 2012.
• On 17 November, the Met Office reported that its Sheffield weather station had recorded its wettest ever autumn.	• Up to 80 million m³ of peat has been removed from the Peak District, removing storage for up to 520 billion litres of water, which instead flows rapidly into rivers, including the River Don.
• Peatlands and moorlands upstream, such as areas of the High Peak, have been drained, so they cannot store water. This means it flows more quickly into the river.	• Building on the floodplain, such as the village of Fishlake, means residential areas are vulnerable to flooding, and increases the impermeable surfaces, so more water flows into the river.

Impacts of flooding

REVISED ○

Impacts	+ / −
Floodwater picks up pollutants like oil or chemicals and takes them downstream	
Spreads diseases	
People and animals can drown	
Provides water for crops	
Roads and railways may be disrupted	
Buildings and things inside them can be damaged	
Provides sediment which makes soil more fertile for crops	
Crops growing on the floodplain may be washed away	
Provides a habitat for fish which people can eat	
Once your house has been flooded, insurance becomes more expensive	

Revision activity

Complete the table opposite to summarise the impacts of flooding. For each impact write + or − next to it to show if it is positive or negative. Colour each statement: either green, if it impacts the environment, or red, if it impacts people.

River management strategies

There are two main responses to flooding: hard engineering and soft engineering.

- **Hard engineering** means strategies to control natural hazards which do not blend into the natural environment – this involves building large structures to try to control the river.
- **Soft engineering** means strategies to control a natural hazard which blend into the environment. These are strategies which attempt to reduce flooding without damaging the river for future generations. This makes it **sustainable** – meaning you can keep on doing it without causing problems to people or the environment in the future.

Revision activity

1 Decide whether each of the methods is an example of hard or soft engineering.

2 Put a tick or cross in the last column to show whether it is sustainable.

3 Make a revision card for each method, giving the correct purpose, type of engineering, whether it is sustainable and why.

Methods	Purpose	Hard or soft engineering?	Sustainable?
Straightening and deepening the river	Water flows faster in a straighter and deeper river so it leaves the area without causing problems		
Building **embankments** or levees	Make the river banks higher so more water can be held in the channel		
Afforestation (planting trees)	Trees take up water by their roots and reduce the amount that reaches the river, so it is less likely to flood		
Washlands	Parts of the floodplain, often used as pasture in the summer, that can be allowed to flood in the winter – one form of flood storage area		
Building **dams**	Allow water to be stored rather than surging downstream and causing floods there. They also can be used for water supply, hydroelectricity and recreation		
Land use zoning	Land that is at the highest risk of flooding is not used for housing but for playing fields or for pasture (from which animals are removed when the flood risk is high)		
Flood walls	Walls built beside rivers that are likely to flood in urban areas – they take up less space than levees		

Case study: A river management scheme – the Mississippi

The Mississippi is the fourth longest river in the world. It is an essential river for the USA, providing 18 million people with their water supply. Flooding is an almost annual event. Severe floods in 2001 caused $13 million of damage and 4400 people had to move.

Hard or soft engineering	Description	Evaluation: is management sustainable?
Hard engineering	Raised embankments called **levees** have been built 15 m high, for 3000 km along the river	Levees are partly blamed for the 2001 floods – they protect the area where they are built but push the problem downstream After a flood, silt is deposited on the channel bed instead of the floodplain, so normal water level rises to higher than the floodplain. Parts of New Orleans lie 4.3 m below river level, increasing the risk of flood damage
Hard engineering	Over 100 **dams** have been built on tributaries	Dams trap silt, preventing it reaching the delta (so birds such as the heron are endangered there) or enriching farmland (so more fertiliser has to be used)
Hard engineering	Engineers cut through meanders to straighten 1750 km of channel, to make the river flow faster	A straightened river loses its variety of habitats for plants, fish and insects The river erodes the banks to resume its natural meandering course, so money and effort have been wasted
Soft engineering	**Afforestation** (tree planting) in the Tennessee Valley – trees absorb water	As well as preventing water from reaching the river, tree planting helps to reduce soil erosion and provides wildlife habitats and opportunities for recreation
Soft engineering	Safe flooding zones – houses near the river are bought and demolished (for example, Rock Island, Illinois) and areas of floodplain are turned into green spaces	It is cheaper in the long term to prevent property damage than to compensate owners when the damage has happened Wetland habitats close to rivers can be preserved

Now test yourself

TESTED ◯

1 Give two hard engineering methods used on the Mississippi.
2 Give two soft engineering methods used on the Mississippi.
3 How high were the levees built along the river?
4 How many dams were built?
5 Give an example of a safe flooding zone.
6 How many kilometres of river were straightened?
7 How far below river level are parts of New Orleans?
8 Name one species of bird endangered by the lack of silt going to the delta.

Revision activity

Draw a spider diagram for the Mississippi. Put each management strategy on a separate leg. Colour hard engineering strategies red and soft engineering strategies green. For each, add a description, including at least one figure or place name. Then give each leg two branches, and use them for the positive and negative aspects of the strategies, making sure you have place names or figures if possible.

Exam practice

1 With reference to a river in the British Isles, explain one physical cause of flooding. [3]

2 Explain how people contributed to flooding in your case study of a river in the British Isles. [3]

3 Choose one hard and one soft engineering method used for managing rivers. Explain how they help prevent floods. [6]

4 Study the photograph below. Explain how flooding can affect people and the environment. [4]

5 For one river outside the British Isles, evaluate the river management strategy used, with reference to the principles of sustainable development. [7]

Revision activity

Make a postcard-sized summary of your case study. Check what you need to know. Make sure you include at least two pieces of detailed information, like place names or numbers, for everything.

Exam tip

For questions like question 2, make sure you give case study detail, and explain in full.

Exam tip

Question 5 needs case study detail, giving good and bad effects of the strategy. At the end you need to give your overall assessment of whether it is sustainable or not.

Coastal processes and landforms

> **You need to be able to:**
> - understand the difference between destructive waves and constructive waves, and how they change the coast
> - understand processes of erosion, transportation and deposition
> - explain the formation of erosional and depositional landforms
> - use aerial photographs and OS maps to identify coastal landforms and land uses.

Waves

REVISED

Coastlines are dynamic – this means that they change through time. Cliffs can crumble and collapse; beaches may have more or less sand this year than last. Wave action is the main process that causes these changes. Destructive waves attack the coast whereas constructive waves add sediment to the shore.

Waves are energy moving through the water. In deep water, the wave creates a circular movement, which just moves the water up and down at the surface. When the wave reaches shallow water, friction with the sea bed slows the wave down. The top of the wave keeps going and falls over the top. This is the wave breaking.

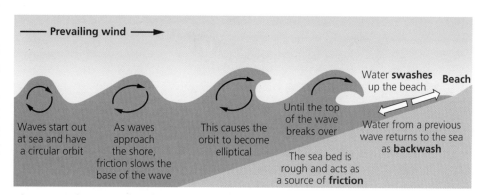

Figure 1 Why waves break

When the wave breaks, the water runs up the beach. This is called swash. It can carry sand or stones from the sea and deposit them on the beach.

When the wave runs back down the beach, it can carry sand or stones out to sea. This is called backwash.

The type and strength of the wave is affected by several factors:
- Faster wind speed makes bigger waves.
- Wind blowing for a longer time makes bigger waves.
- Wind blowing over a longer stretch of water makes bigger waves. The distance is called the fetch.

If a strong wind blows 5000 km across the Atlantic ocean for 5 days, it will make bigger, more powerful waves than a gentle breeze blowing 150 km across the Irish Sea for 3 hours.

We can divide waves into constructive and destructive waves.

	Destructive waves	Constructive waves
Size	High and close together	Low and far apart
Frequency	Frequent, up to 15 per minute	Less frequent, 6–9 per minute
Season	Common in winter (storm waves)	Common in summer
Effects	Stronger backwash than swash. Drags sand and pebbles out to sea. Erodes the coast	Strong swash and weak backwash. Pushes sand and pebbles up the beach. Causes deposition which builds up the coast

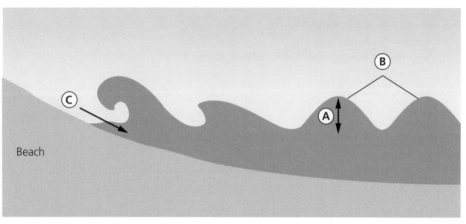

Figure 2 Destructive waves.

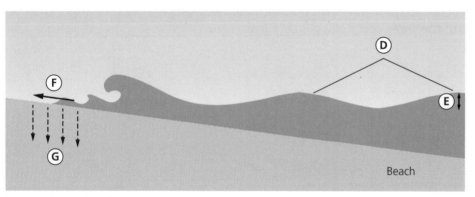

Figure 3 Constructive waves.

Now test yourself   TESTED ◯

1   Figures 1 and 2 show destructive and constructive waves. Match the labels below
    with the letters A–G on the diagrams.
    ● higher wave
    ● waves far apart
    ● water sinks into beach reducing backwash
    ● strong backwash pulls sediment down the beach
    ● lower wave
    ● waves close together
    ● strong swash pushes sediment up beach.
2   Which type of waves have stronger backwash than swash?

Coastal erosion, transportation and deposition

Erosion processes

The sea can break up rock in four ways. These are similar to the ways rivers break up rocks, but sometimes different names are used.

- **Hydraulic action** – this is the force of the water wearing away the rock. If there is a crack in the rock, the water can compress air into the crack, which puts pressure on it and makes the crack bigger.
- Abrasion (corrasion) is the sandpapering action of water carrying sand and pebbles – as the sand and pebbles are moved along against the rock, they scrape the surface away like sandpaper, which smooths and wears away the rocks where waves hit at the base of a cliff.
- Solution (corrosion) is the chemical action of seawater dissolving minerals in rocks such as limestone. This weakens the rock and makes it break up more easily.
- Attrition occurs as pebbles transported by waves hit against each other. They break into smaller particles, and wear away the jagged edges, becoming rounder.

Exam tip

Whenever you write about one of these processes, try to use the name **and** explain what it is. And be careful about corrasion and corrosion! Remember corrasion rhymes with abrasion, so it's easy to remember they are the same thing.

(a)

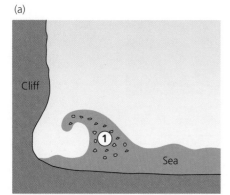

(b)

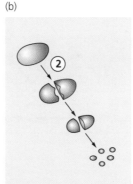

(c)

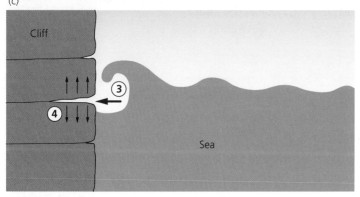

Figure 4 Coastal erosion processes.

Now test yourself

1 Identify the three erosion processes shown in Figure 4a–c.
2 Draw quick sketches of the three diagrams and select appropriate labels for 1–4 from the list below.
   a) Air compressed (squashed) by advancing wave.
   b) Wave carries sand and pebbles which wear away the rock.
   c) Compressed air expands after wave breaks and loosens blocks of rock which fall into the sea.
   d) Pebbles crash against each other becoming smaller and rounder.

27

Transportation

The pieces of rock eroded by the sea are moved by wave action. This is called **transportation**. The sea can carry particles along using the same processes as a river.

When waves break normally on a beach, each swash carries this material up the beach and the backwash sweeps it back out to sea, as shown in Figure 5a.

However, there are lots of places where the wind blows the waves towards the beach at an angle. When this happens, the swash carries material up the beach at the same angle, but the backwash returns it straight back to the water's edge because of gravity (see Figure 5b). This zigzag movement of sand or pebbles along the coast is called longshore drift.

This name should be quite easy to remember, as it results in the particles **drifting along** the **shore**.

Revision activity

Try to name and explain the four main transportation processes from the rivers section. If you can't remember, check back to page 17 and then try again.

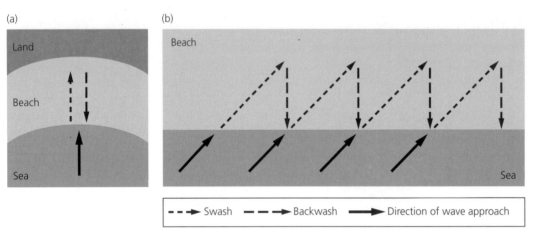

Figure 5 Coastal transportation.

Deposition

When waves are constructive they deposit the load they carry onto the shore. This is called **deposition**. This often happens in a sheltered bay, between two headlands.

Now test yourself   TESTED ◯

What coastal transport process involves pebbles moving in a zigzag fashion along a beach?

Revision activity

Make a list of the words you need to learn in this section. Look at the letters they start with, and make up a silly sentence where the words start with those same letters. The sillier the sentence, the better you will remember it! For instance:
- hydraulic action   hedgehogs
- attrition   are
- abrasion   all
- solution   spiky

Exam practice answers and quick quizzes at **www.hoddereducation.co.uk/myrevisionnotesdownloads**

Landforms resulting from erosion and deposition

Landforms are recognisable features in the landscape, created by natural processes:

- **Erosional landforms**: headland, cliff, wave cut platform, cave, arch and stack.
- **Depositional landforms**: sandy beach, shingle beach, and spit including hooked spits.

Erosional landforms

Where the coastline has areas of harder and softer rock, the soft rock will wear away, leaving hard rock sticking out to sea. This makes a headland. When destructive waves approach this rocky headland, the processes of hydraulic action and abrasion erode a wave cut notch. Through time, this notch gets bigger and the unsupported rock above it is undercut and collapses into the sea, forming a cliff. Continued erosion and repeated collapse causes the cliff line to retreat. This leaves an almost level area of rocks and rockpools known as a wave cut platform at the base of the cliff.

You can see a wave cut platform at Portstewart as you walk along the cliff path towards the Strand.

Erosion attacks the rock where it is weak, to form caves. Hydraulic action is very effective in making caves bigger as air gets trapped in the cave by advancing waves and is forced into cracks in the roof and at the back of the cave. Over time this means the cave can extend right through the headland to form an arch. When the roof of the arch collapses it leaves an isolated stack which can be eroded to leave only a stump.

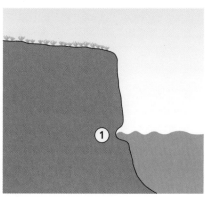
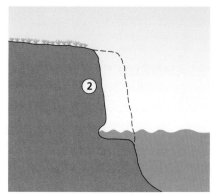
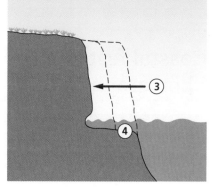

Figure 6 Cliff retreat.

Figure 7 Caves, arches and stacks.

Theme B Coastal Environments

29

Deposition landforms

A **beach** forms where sand or shingle (pebbles) builds up in a sheltered position on a coastline. It is built by constructive waves which, over time, add sediment with their swash – the backwash is too weak to remove all of it. On most beaches the swash and backwash both operate straight up and down the beach so sand or pebbles do *not* move along the shore (see Figure 5a, page 28). Shingle beaches are usually steeper than sandy beaches. This is because the pebbles let the water through easily, so there is very little backwash, so pebbles are not pulled back down the beach.

A **spit** is an extended beach that forms a 'finger' of land sticking out into the sea. It is formed when sediment is carried along by longshore drift and deposited out into the sea where the coastline turns a corner. Spurn Head in England is a classic example of a spit.

Figure 8 The formation of a spit.

Now test yourself   TESTED ◯

1   For each of the numbers 1–5 in Figure 8, select an appropriate label from the list below to explain the formation of a spit.
   (a) Deposition in sheltered area behind spit.
   (b) Sediment moves along the beach by longshore drift.
   (c) Spit grows as sediment is added, continuing in a straight line even though the coast changes direction.
   (d) Waves approach coast at an angle because of the prevailing (most common) wind direction.
   (e) A second common wind direction or river flow may make the end of the spit curve round, creating a **hooked spit**.

2   In the list Cliff, Arch, Cave, Spit, Stack, which **coastal landform** is the odd one out and why?

Coastal features and land uses on aerial photographs and OS maps   REVISED ◯

Now test yourself   TESTED ◯

1   Using the OS map on the inside back cover and the guidance on page 11, match the grid references a–d with the correct coastal features (1–4).

Grid references	Coastal features
a 242257	**1** Cave
b 246286	**2** Arch
c 2426	**3** Beach
d 2532	**4** Cliffs

2   Study the photo of Spurn Head below. In what direction must sediment be moving by longshore drift to form and extend this spit?

Exam practice answers and quick quizzes at **www.hoddereducation.co.uk/myrevisionnotesdownloads**

Revision activity

Practise drawing your own diagrams of erosional and depositional coastal landforms and giving them labels in full sentences to explain how they are made.

Exam practice

1 Give two factors which lead to bigger waves. [2]

2 Tick the correct box to show which type of wave has each characteristic. [3]

Constructive waves	Characteristic	Destructive waves
	High	
	Weak backwash	
	Remove lots of material from the beach	

3 Explain how abrasion can erode rock at the coast. [3]

4 Explain how a wave cut platform is formed. [5]

5 Explain how longshore drift transports material along a beach. [4]

6 Name two coastal landforms formed by deposition. [2]

7 Study the photograph below.

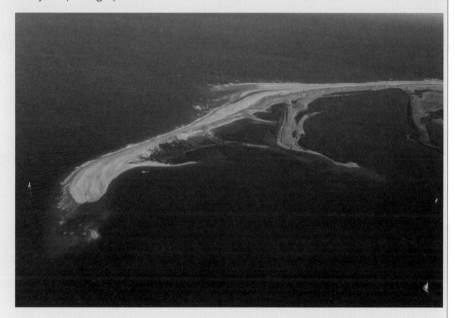

Name one feature formed by deposition and explain how it is made.

Feature _____ [1]

How it is made _____ [4]

8 Using the map on the inside back cover, find two pieces of map evidence that the coast is used for tourism. [4]

Evidence	Grid square

Sustainable management of coasts

- understand why coastal defences may be needed
- describe and evaluate hard and soft engineering strategies for coastal management
- evaluate the management of a coast in the British Isles to decide if it is sustainable.

Reasons for coastal defences

REVISED ●

Coastal defences are ways of trying to stop the sea from eroding the land. People have been trying to do this for many years, because people use the coastal zone so much.

There are three main reasons why coastal defences may be needed:
1 **People live near the coast**: just over half the world's population – around 3.2 billion people – live within 200 km of the sea. In all continents, except Africa, the majority of people live near coasts.
2 **Economic importance**: people earn money near coasts by using ports to send their goods abroad, fishing and tourism. Tourism generates 60% of Majorca's gross national product (GNP), mostly based around the beaches.
3 **Sea level is predicted to rise by 0.5–1 m as climate change continues**. This means even more places will be affected by the sea. Some places could be flooded, and other places will have more storms, which means more erosion by the waves. The residents of the Pacific island of Tuvalu have already made plans to leave.

Exam tip

Try to learn a figure to go with each of these reasons, so you can include it in your answer.

Some people say this means we need more coastal defences. Other people say we should move away from the sea, and let the sea erode or deposit material naturally.

Now test yourself

TESTED ●

What is the connection between global warming and coastal defences?

Coastal management strategies

Coastal management means making decisions and taking action to control the land uses and natural processes happening at the coast. For example, if a building is at risk of falling into the sea because the sea erodes the land it is standing on, the council could decide to manage this by using coastal defences to try to stop the erosion. This would be their coastal management strategy.

1 **Hard engineering methods** – these involve building large structures, usually of concrete or wood.

- Sea walls have been used for a long time to stop the sea. They look like a concrete wall at the back of a beach. Some sea walls are curved, others have steps, and others are straight. They are designed to stop the waves coming over the top and doing damage during storms. Curved walls are supposed to send the wave's energy back into the sea, and steps are supposed to help break up the wave's energy, so both are designed to stop erosion. These are expensive (£10 million per kilometre) but may be worth the money to protect an area like a town with lots of people and property.

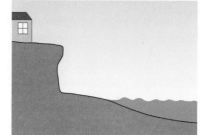

- Groynes – these are long fences made of heavy wood, stretching out from the beach into the sea. They are designed to trap sand moving along by longshore drift. This creates a wide beach where the groynes are, so waves use up their energy crossing it and cause less erosion behind the beach. These are relatively cheap at £5000 per metre. However, they only last twenty years. They trap sand which naturally would travel further along and protect another part of the coast, so they may cause erosion problems elsewhere.

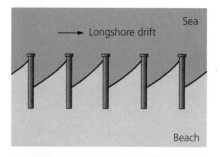

- Gabions – these are wire boxes filled with stones, often put at the bottom of cliffs. They are designed to absorb and break up the waves' energy, and stop it eroding the cliff. They are cheap, but the metal rusts and can be broken in severe storms, so they do not last long.

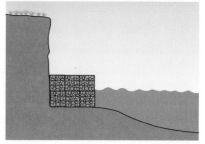

> **Exam tip**
>
> In an exam, you could be asked to evaluate the methods. Make sure you can give both problems and benefits, and then an overall opinion – do you think they are a good idea?

2 **Soft engineering methods** – these involve working with nature in a way that fits in with the environment so it causes less damage.

- Beach nourishment – this is where large amounts of sand are brought onto the beach from somewhere else and deposited to help make a wider or deeper beach to protect the land behind it by absorbing the waves' energy. This is cheap (£3000 per metre) but needs lots of maintenance, as the sand will be eroded away.
- Managed retreat – this is where people decide to move their land uses further away from the coast so they cannot be eroded. This may mean paying people compensation to move inland, and allowing their house to fall into the sea. This may upset people, but is a more long-term solution.

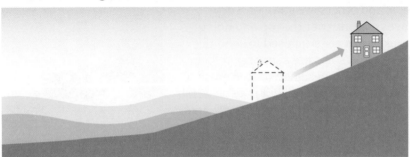

Revision activity

Make yourself a table like the one here and complete it to show the problems and benefits of the coastal management methods. Colour-code hard and soft engineering.

Strategy	Problems	Benefits
Sea walls		
Groynes		
Gabions		
Beach nourishment		
Managed retreat		

Now test yourself TESTED ◯

1 How are sea walls shaped:
  a) to help break up wave energy?
  b) to reflect wave energy back out to sea?
2 What term means adding sand or shingle to a beach?
3 What strategy is often used to retain (keep) sand or shingle on a coast where longshore drift operates?

Newcastle in County Down has had several coastal management strategies.

You need to describe them and evaluate whether they have been sustainable.

Strategy	Impact	Evaluation: + or -
Groynes – concrete groynes were built on the beach in the 1980s	Trapped sand which was moving north east	+ This helped protect the beach, attracting more tourists to the area, bringing increased income
	Lasted about twenty years, then decayed and were useless	− Replacing groynes is expensive (£5000 per metre)
Gabions – new wire mesh boxes filled with local stone were placed near the mouth of the River Shimna in 2006	Protect recreation ground	+ This was successful in protecting a valuable tourist attraction and therefore protecting income from tourism
	Protects the footbridge, gives better tourist access to the promenade	+ Gabions seem to have been successful at breaking up wave energy
	Earlier gabions decayed over time and had to be replaced	+ Regular replacements cost a lot of money
Sea wall – the original Victorian sea wall was badly damaged in a storm in 2002 and replaced	The original wall protected buildings for many years	+ This allowed economic development, bringing in income
	Storm damage meant a new wall was needed	− This cost £4m, which is only sustainable if it protects valuable property
	The curved wall means reflected waves are increasing erosion of the beach on their way back down	− This means the wall may be unsustainable as it may remove part of the beach, making overall erosion worse and removing a tourist attraction

Overall, it appears that the coastal management strategies, all hard engineering, are all unsustainable, as they can cause further problems and they require significant maintenance.

Revision activity

You can use this table to help write an answer to an exam question. Try the following question – the first part is done for you.

'Evaluate the sustainability of the coastal management strategy used to protect a named coastline in the British Isles.'

In Newcastle, County Down, several hard engineering strategies have been used.

Concrete groynes were built on the beach in the 1980s. These trapped sand which was moving north east. This was positive because *it helped protect the beach, attracting more tourists to the area, bringing increased income.*

However, they only lasted about twenty years, then decayed and were useless. This was negative *because replacing groynes is expensive (£5000 per metre).*

You need to do the same thing for gabions and the sea wall.

At the very end you need to give an overall evaluation. There is information under the table to help you!

Exam tip

You need to take information from the table and bring it together – check the sample answer and see how it uses the information from the table. You also need to add in an evaluation – say whether it is positive or negative.

Exam tip

Bear in mind that if a question asks you to evaluate, you need to give both the good and the bad, and give an overall decision at the end.

Theme B Coastal Environments

Now test yourself

TESTED

1. Name two strategies, other than sea walls, used to protect the coast at Newcastle, Co. Down.
2. For each strategy in the table on page 35, write down whether it was sustainable overall, and why.

Exam practice

1. Tourism accounts for 60% of Majorca's income. Explain how this might make coastal defences necessary. [3]

2. Study the photograph below showing a method of coastal management. Give the name and state whether it is hard or soft engineering.

Name _____

Hard or soft engineering _____ [2]

3. Describe and evaluate how sea walls help to defend the coastline. [8]

4. State one advantage and one disadvantage of beach nourishment. [2]

5. With reference to a case study of coastal management in the British Isles, describe and evaluate the strategy used, referring to the principles of sustainable development. [9]

Theme C Our Changing Weather and Climate

Measuring the elements of the weather

You need to be able to:

● distinguish between weather and climate
● describe how elements of the weather are measured
● know where forecasters get their information.

Distinguish between weather and climate

REVISED

Weather is the day-to-day condition of the atmosphere, such as how hot it is, or whether it is raining. Climate is the average weather taken over about 35 years – the sort of weather we usually expect to get.

Now test yourself

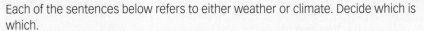TESTED

Each of the sentences below refers to either weather or climate. Decide which is which.

(a) The day of the wedding turned out to be beautiful, with clear blue skies.

(b) Daisy's trampoline blew halfway across the garden in a storm.

(c) When she planned a picnic in November everyone said Brenda was crazy.

(d) In September, all the shops start to sell gloves, hats and scarves.

(e) Oonagh's mobile phone was ruined when she got caught in a rainstorm.

(f) The farmer in Spain had lots of irrigation to provide water for his crops, because he knew there wouldn't be enough rain.

Exam tip

Listen for the sounds to help you remember. Is it about **w**eather at **o**ne time? Then it's **w**eather. Is it about what the weather is **c**ommonly like? Then it's **c**limate.

Measuring the elements of the weather

The **elements** of weather are the different things which make up weather. These are shown in Figure 1.

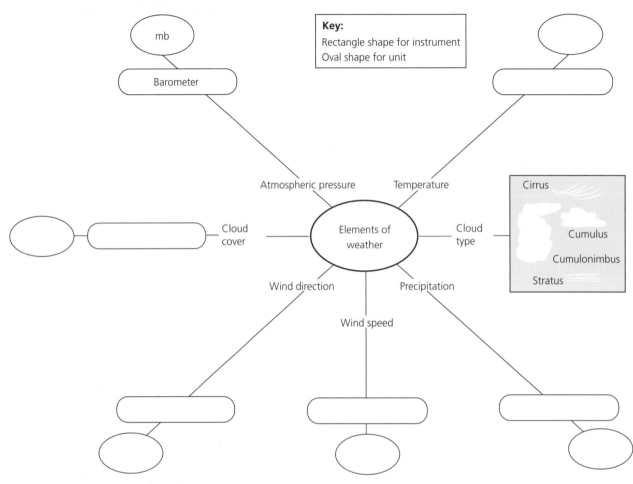

Figure 1 The elements of weather.

Complete Figure 1 by selecting an instrument and appropriate units from the two lists. 'Atmospheric pressure' has been completed for you.

Instruments:
- Rain gauge
- **Barometer**
- Wind vane
- Digital thermometer
- Anemometer
- Observation

Units:
- Degrees Celsius (°C)
- Millimetres (mm)
- **Knots**
- **Millibars** (mb)
- Eight points of the compass (N, NE, E, SE, S, SW, W, NW)
- **Oktas**

How to measure the weather

Precipitation	A **rain gauge** is placed in the open so there is nothing stopping the rain. Read the level of water in the measuring cylinder of the rain gauge at the same time each day. Then empty it and reposition the rain gauge ready to record the rainfall over the next 24 hours
Temperature	Temperature readings are taken with a **digital thermometer**, which uses a simple metal sensor and displays the temperature digitally. It includes a memory feature which may record highest and lowest temperatures that day. Thermometers are kept in a Stevenson screen, which is white to reflect heat, provides shade, but has vents in the sides to allow the air to flow through to give a fair measurement of temperature
Wind speed	An **anemometer** usually has small cups which spin round as they catch the wind. Most anemometers have digital readouts which tell you the wind speed. An anemometer is kept high up so nothing can shelter it from the wind
Wind direction	On a **wind vane**, the arrow points to where the wind is coming from. A wind vane is also kept high up, such as on a high roof, so nothing can shelter it from the wind
Cloud cover	Look at the sky and estimate how many eighths (oktas) of the sky are covered in cloud
Cloud type	Look at the sky and identify the type of cloud – you should be able to recognise **cumulus**, **cumulonimbus**, **cirrus** and **stratus** (check the sketch in Figure 1)
Atmospheric pressure	Read the needle on a barometer. There may be a second needle – this can be set to show what the pressure was at any one time, and the main needle then shows how pressure has changed

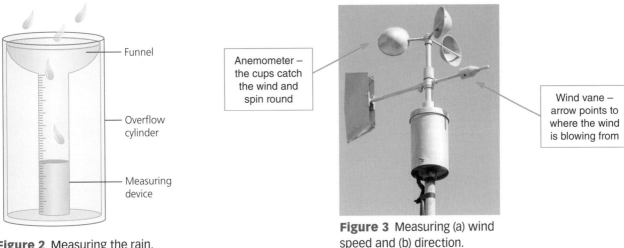

Funnel

Overflow cylinder

Measuring device

Figure 2 Measuring the rain.

Anemometer – the cups catch the wind and spin round

Wind vane – arrow points to where the wind is blowing from

Figure 3 Measuring (a) wind speed and (b) direction.

Sources of data for weather forecasting

REVISED ●

Forecasters used to rely on pieces of seaweed or the colour of the sky to help them create a weather forecast. Now they have a lot more data.

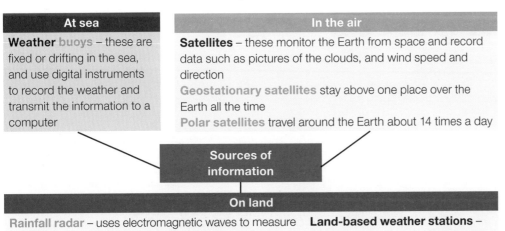

At sea
Weather buoys – these are fixed or drifting in the sea, and use digital instruments to record the weather and transmit the information to a computer

In the air
Satellites – these monitor the Earth from space and record data such as pictures of the clouds, and wind speed and direction **Geostationary satellites** stay above one place over the Earth all the time **Polar satellites** travel around the Earth about 14 times a day

Sources of information

On land

Rainfall radar – uses electromagnetic waves to measure where rain is happening and how heavy it is at that time. This can be shown on a map, and combined with what forecasters know about how weather patterns usually move to help predict where rain will happen later on

Land-based weather stations – every three hours these record all the elements of the weather discussed in the last section. There are over 10,000 of these around the world

Figure 4 Sources of weather information.

Now test yourself

1 Name the instrument used to measure wind speed.
2 Name the instrument used to measure precipitation.
3 What are the units of measurement of pressure?
4 What are the units of measurement of temperature?
5 What element of weather is measured in oktas?
6 What is the white wooden box where thermometers are located called?
7 Why is this the most suitable place for measuring air temperature?
8 Where should a rain gauge be placed for accurate readings?
9 Name three possible sources of data used to create a weather forecast.

Revision activity

Search for an image of each weather instrument. Print these images and label them in felt tip with their name, what they measure, how they work and the unit of measurement. Stick them on the wall until you know them really well.

Exam practice

1 Define climate. [2]
2 Explain how you would keep an accurate record of daily precipitation. [3]
3 Study the photo below which shows a Stevenson screen. Explain two reasons why it is used to store thermometers. [2]

4 Study the rainfall radar map below. Explain how this could help with weather forecasting. [3]

Key
Rainfall, mm/h
0.5–1.0
2.0–4.0
8.0–16.0

Exam practice answers and quick quizzes at **www.hoddereducation.co.uk/myrevisionnotesdownloads**

Factors affecting climate

You need to be able to:

● explain how climate is affected by latitude, prevailing winds, distance from the sea and altitude.

Different parts of the world have different climates. These are affected by four main factors: latitude, prevailing winds, distance from the sea and altitude.

Latitude

REVISED ◯

Latitude means how far north or south of the equator a place is. The equator has a latitude of 0°, and the North Pole is at latitude 90°.

The Sun shines directly overhead at the equator, so its rays are concentrated in a small area which gets heated up. Near the poles, the same amount of solar radiation is spread out over a larger area, so it does not heat up as much.

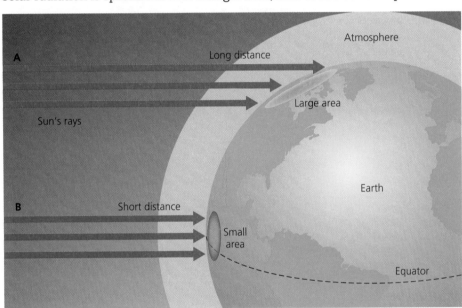

Figure 5 Where the Sun's rays meet the Earth at the equator and the poles.

Exam tip

Latitude is usually the main factor affecting the global pattern of climate. All the others just change small areas.

Revision activity

To help you remember how and why latitude affects climate, you can see it for yourself with a torch.

Shine the torch directly down on to a table. Then shine the torch at an angle onto the table. Compare the shape made by the light, and how bright it is in each case.

You should find that the light is brightest when the torch shines directly down, because it is concentrated in a small area.

Prevailing winds

REVISED ◯

The prevailing wind is the most common direction the wind blows from.

If the wind blows from a warm place, it will bring warm air. If the wind blows from a cold place, it will bring cold air.

If the wind comes over the sea, it will pick up water and bring rain. If the wind comes over land, it will be dry.

Revision activity

To help you remember this, go to the kitchen while the oven is on. Feel the heat from the oven. Then open the fridge door and feel the cold air. Imagine wind blowing from them.

41

Distance from the sea

In summer, the land heats up more quickly than the sea. Places close to the sea will be kept cool, but places in the middle of a continent can get very hot.

In winter, the land loses its heat more quickly than the sea. Places close to the sea will be kept warmer, but places in the middle of a continent can get very cold.

Belfast is close to the Atlantic Ocean, and Moscow is in the middle of a continent. Even though Belfast and Moscow are almost the same distance from the equator (55°N), Belfast stays cool in the summer (14°C) while Moscow can get very hot (average 23°C). In winter, Moscow's average temperature is –7.5°C, while in Belfast the average is 4°C.

> **Exam tip**
>
> **C**ontinents have **c**olossal temperature range – they **c**an be **c**old or **c**ooking!
>
> **M**aritime (near the sea) climates are **m**oderate – **m**ild winters, **m**ild summers.

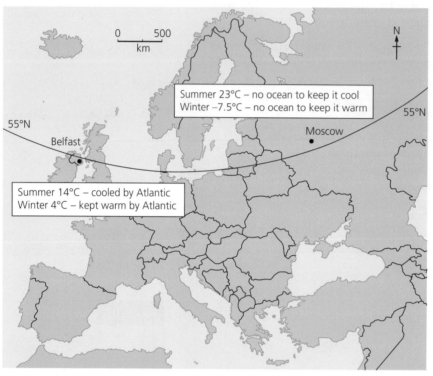

Figure 6 How distance from the sea affects climate.

To summarise:
- Places near the sea are always milder – not as hot in summer, not as cold in winter – than places in the middle of a continent.
- Places in the middle of a continent are drier.
- Places near the sea have more rain.

Altitude

Altitude means how high the land is above sea level. As you climb a mountain it gets colder, about 1°C every 100 m. This is because the Sun heats the ground and the ground heats the air. As you go higher, there is less ground to heat the air. Also, the air is at lower pressure, so it spreads out and loses heat.

> **Exam tip**
>
> You need to be able to use the technical words and explain what they mean. They are likely to give you a real place and ask you to explain why it is hot/cold/wet/dry based on information like its latitude or how close it is to the sea, so make sure you really understand these ideas.

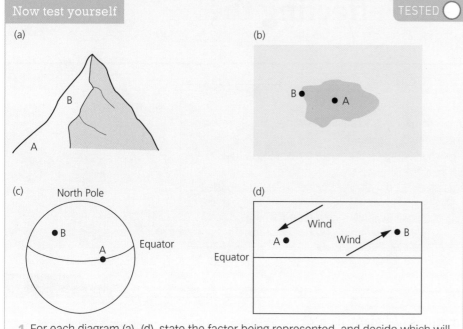

(a)

(b)

(c) North Pole

(d)

1 For each diagram (a)–(d), state the factor being represented, and decide which will be warmer in each case, A or B.

2 What do we call the distance from the equator?

3 Why is it hotter near the equator?

4 What do we call the most common wind direction?

5 Which heats up faster, land or sea?

6 Which cools down quicker, land or sea?

7 Why is Moscow hotter in summer than Belfast?

8 What do we call the height of the land?

9 If the temperature is 15°C at sea level, what temperature will it be 500 m up a mountain?

10 Which place will be wetter: Moscow or Belfast?

Exam practice

1 Explain how latitude affects climate. [4]

2 Explain why two places at the same latitude may have different climates. [4]

3 Study the diagram below which shows temperatures at different altitudes on a mountain.

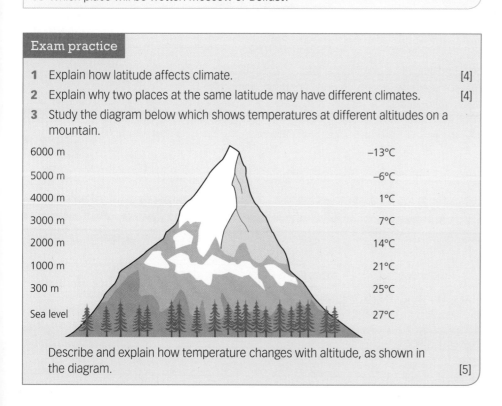

6000 m	−13°C
5000 m	−6°C
4000 m	1°C
3000 m	7°C
2000 m	14°C
1000 m	21°C
300 m	25°C
Sea level	27°C

Describe and explain how temperature changes with altitude, as shown in the diagram. [5]

43

Weather systems affecting the British Isles

- understand the temperature and moisture characteristics in winter and summer of the four main air masses that affect the British Isles
- understand the weather patterns and sequence of change associated with a frontal depression as it moves across the British Isles
- understand the weather patterns associated with anticyclones in the British Isles during winter and summer
- interpret synoptic charts and satellite images and understand the limitations of forecasting.

Air masses

REVISED

An air mass is a large body of air (think of a giant 'blob' of air, perhaps the size of western Europe). An air mass will take on the characteristics of its location. This means that each air mass blowing towards the British Isles will bring a particular type of weather.

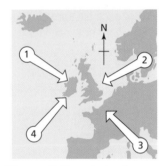

Figure 7 Four main air masses affecting the British Isles.

Name	Blows from	Characteristic
Maritime	Ocean	Moist – brings rainfall
Continental	Continent	Dry
Polar	Somewhere near the North Pole	Cold in winter
Tropical	Somewhere near the tropics	Hot in summer, mild in winter

There are four main air masses which affect the British Isles, which can be seen in Figure 7. These air masses get their characteristics from the surface they blow over.

Exam tip

There is one air mass which can be surprising. Polar continental is very cold in winter. However, in summer, the middle of a continent warms up, so polar continental in summer can be quite warm.

Now test yourself

TESTED

Complete the table below by selecting the correct descriptions from the table in columns B and C.

| Label on Figure 7 | A: air mass name | B: temperature characteristics | | C: moisture characteristics |
		Summer	Winter	
1	Polar maritime	Warm/cool	Warm/cold	Wet/dry
2	Polar continental	Hot/cool	Warm/cold	Wet/dry
3	Tropical continental	Hot/cold	Mild*/cold	Wet/dry
4	Tropical maritime	Warm/cold	Mild*/cold	Wet/dry

* Mild means not very cold but not warm either.

Weather patterns associated with depressions and anticyclones

> **Exam tip**
>
> You need to know what weather occurs in different parts of a **depression** – at the **warm front**, in the **warm sector** and at the **cold front**, and you need to be able to talk about places.

Weather systems

Anticyclones and frontal depressions are weather systems. This means they always act in a fairly predictable way, which determines the weather of the British Isles and allows us to make weather forecasts.

Figure 8 Comparison of anticyclones and depressions.

	Anticyclone	Depression
Definition	An area of **high pressure**	An area of **low pressure** *(easy to remember – if you're depressed, you're feeling 'low')*
Air movement	Air is **sinking**	Air is **rising**
Cloud cover	**Clouds cannot form**	As it rises, air cools and condenses to form **clouds**
Wind speed	Winds are **gentle**, blowing **out** from the centre of high pressure. The isobars are far apart	Winds are **strong**, blowing **into** the centre of low pressure. The isobars are close together
Wind direction	Winds generally blow **clockwise** *(remember – **anti**cyclone and **anti**clockwise do not belong together)*	Winds generally blow **anticlockwise**
Duration	Anticyclones are slow-moving so weather **remains the same** for several days or even a week or more	A depression moves quickly so the whole depression can pass over a particular location in around **24 hours**
Weather	<ul><li>Always **dry; little or no cloud; calm or gentle wind**</li><li>In **summer**, day temperatures are **high** (because of cloudless skies), night temperature can be **cool** as the Earth's heat radiates into the atmosphere</li><li>In winter, day temperature is **low** because days are short and Sun is not powerful. At night, **cloudless skies** allow the temperature to fall below 0°C and **frost** forms. Air near the ground is chilled and any moisture in it condenses to form **fog**</li></ul>	A **predictable sequence of weather** occurs as the depression moves over any location 1 First the warm **front** approaches. At the front, warm air rises over cold air so clouds form and rain falls 2 Between the fronts is the warm sector, so temperatures rise a little, heavy rain stops and there may be some drizzle 3 The cold front arrives. Again the warm, moist air is lifted over the cold air so clouds and heavy rain occur 4 Following the fronts, there is the cold sector, with lower temperatures. Cloud breaks up and showers become fewer

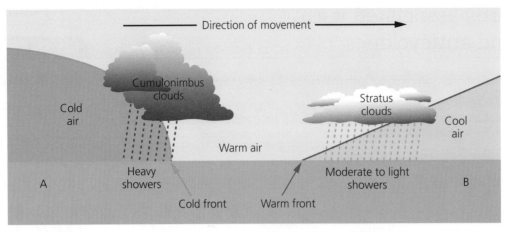

Figure 9 A diagram of a depression.

Revision activity

1 In Figure 8 (on page 45), each row of information follows logically from the row above. Make sure that you can follow the logic and then draw out a simpler version of the table to show the key points, which are printed in bold. Remember that, in almost every way, the characteristics of a depression and an anticyclone are the opposite of each other.

2 In the table below, identify which type of weather system would lead to each impact.

Impacts	Depression, summer anticyclone or winter anticyclone
1 Trees blown down, blocking roads	
2 Drought creates problems for farmers and gardeners	
3 Sequence of rain, showers and bright intervals ensures that crops have sufficient moisture	
4 Fog affects drivers on the M1 and causes delays for aircraft at Belfast International Airport	
5 Heavy rain may cause flooding, for example, in the Somerset Levels	
6 Icy, slippery footpaths cause injuries to elderly pedestrians with increased admissions to the Royal Victoria Hospital	
7 TV weather forecasts advise the use of raincoats and umbrellas	
8 Lots of sunshine ripens crops in East Anglia	
9 Increased sales of ice-cream and suntan lotion	

Now test yourself

TESTED ⬤

1 What are the name and characteristics of the air mass that influences the British Isles when winds are blowing from the north-west?

2 Does an anticyclone have (a) high or low pressure, and (b) rising or sinking air?

3 Where is rain to be expected in a depression: (a) the cold sector, (b) the warm and cold fronts, or (c) the warm sector?

4 Which weather system, depression or anticyclone, brings strong winds?

5 Which weather system is responsible for fog and frost in winter?

Forecasting: synoptic charts and satellite images

Synoptic charts

Synoptic charts are weather maps that summarise the weather at a particular time. The date and time of day are clearly stated and should be noted as they help with the interpretation. A depression will have **fronts** (which are the boundaries between air masses) known as warm, cold and possibly occluded (see Figure 10) and the isobars are close together with the **lowest** value in the centre. Anticyclones have isobars that are further apart and the **highest** value is in the centre.

> **Exam tip**
>
> You could have to interpret synoptic charts and satellite images. A key is always provided with a synoptic chart so you don't have to learn all the symbols.

ᴖᴖᴖᴖᴖ	▲▲▲▲▲	ᴖ▲ᴖ▲ᴖ▲
Warm front	**Cold front**	**Occluded front**
Semi-circles like drawings of the Sun reminding you of warmth. Always moves from west towards east	Shapes like jagged teeth reminding you of a cold 'biting' wind. Always moves from west towards east	This symbol is a mixture of both shapes. Found where warm and cold fronts meet

Figure 10 Symbols on a synoptic chart.

Symbol	Precipitation	Symbol	Cloud cover	Symbol	Wind speed
🌢	Drizzle	◯	Clear sky	◎	Calm
●	Rain	◐	One okta	◯—	1–2 knots
∴	Heavy rain	◔	Two oktas	◯—‚	5 knots
✶	Snow	◔	Three oktas	◯——	10 knots
=	Mist	◑	Four oktas	◯——‚	15 knots
≡	Fog	◕	Five oktas	◯——‚	20 knots
⌐	Thunderstorm	◕	Six oktas	◯——▾	50 knots or more
		◖	Seven oktas		
		●	Eight oktas		
		⊗	Sky obscured		

Figure 12 Synoptic chart symbols

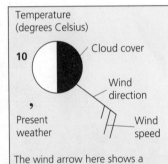

Temperature (degrees Celsius)

The wind arrow here shows a south-east wind. Winds are named according to where they come *from*, just as we call a person French if they come *from* France.

The present weather is drizzle. The symbol can be found on the key provided with each synoptic chart.

Figure 11 Weather at a weather station.

Satellite images

Satellite images are photographs taken from space and sent back to Earth. On a satellite image, a depression will show up as swirls of white cloud along the fronts, on a dark background. An anticyclone will be shown as clear skies, allowing the land and its coastline to be visible.

Centre of the depression, lowest pressure

Swirl of cloud shows lots of cloud and rain at warm and cold front

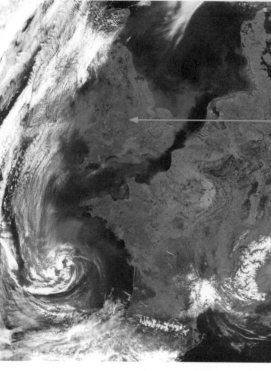

Clear skies mean you can see the outline of the land easily

Now test yourself

1 Which weather system, depression or anticyclone, is represented on a satellite image by a swirl of clouds on a dark background?
2 On a synoptic chart, what type of front has the symbol of a line of jagged 'teeth'?

TESTED ◯

Weather forecasts and their limitations

A weather forecast is a prediction of the weather expected in an area.

Range means how far in advance the forecast is given. It may be for the next 24 hours (short range), the next five days (medium range) or the next three months (long range).

Accuracy means how accurate the forecast turns out to be. A long-range forecast is less accurate than the forecast for the next 24 hours, because we can measure what the weather is doing now and use computer models to help predict what it is likely to do in the near future, but trying to look further ahead involves too many possibilities.

Now test yourself

TESTED ◯

1 What is meant by a long-range weather forecast?
2 What is meant by a short-range weather forecast?
3 Which type is more likely to be accurate?

Exam tip

You could be asked about the limitations of forecasts – these are range and accuracy.

Exam practice answers and quick quizzes at **www.hoddereducation.co.uk/myrevisionnotesdownloads**

Exam practice

1 Complete the table below to show some of the four main air masses affecting the British Isles. [4]

Air mass	Direction it comes from	Summer	Winter
Polar maritime		Cool and wet	Cold and wet
	South-west	Warm and wet	Mild and wet
Tropical continental	South-east		

2 Study the satellite image below showing a weather system over the British Isles in January. Name the weather system and describe the likely weather. [4]

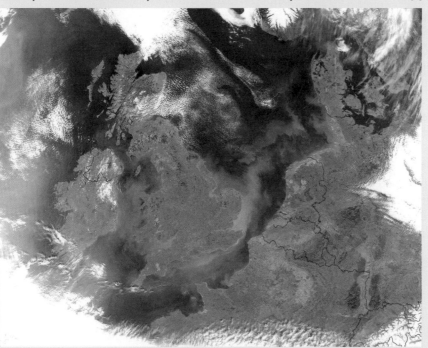

3 As a depression passes from west to east over the British Isles, the weather changes. Describe the three main phases of weather brought by a depression. [6]

4 Study the weather forecast below. Explain what its limitations might be. [4]

London (Greater London)

Today
26° 15°
Sunny.

Sunrise: 05:02   Sunset: 21:11

VH UV   L Pollution   VH Pollen

	Sat 17 Jul	Sun 18 Jul	Mon 19 Jul	Tue 20 Jul
	☀ 28° 18°	☀ 29° 18°	⛅ 26° 16°	☁ 25° 16°

TODAY

Now	12:00	13:00	14:00	15:00	16:00	17:00	18:00	19:00	20:00	21:00	22:00	23:00
☀	☀	☀	☀	☀	☀	☀	☀	☀	☀	☀	☾	☾

Chance of precipitation

<5%	<5%	<5%	<5%	<5%	<5%	<5%	<5%	<5%	<5%	<5%	<5%	<5%

Temperature (°C)

19°	21°	23°	24°	25°	26°	26°	26°	25°	24°	22°	21°	20°

The impacts of extreme weather

● describe the impacts of extreme weather on people and property, using one case study from outside the British Isles. This may be a tornado **or** hurricane **or** drought.

Extreme weather

REVISED ●

Extreme weather means weather that is different from the normal weather we experience. The UK has very little extreme weather, apart from a few storms and occasional heatwaves. Other parts of the world have far more extreme weather, which can have very severe impacts on people:

● **Drought** – this is a long period of time with significantly less rain than usual. This can result in crops failing and famines.
● **Tornadoes** – these are columns of air which rotate violently with winds so strong they can pick up large items such as lorries. At their most extreme the winds are more than 300 mph. They destroy buildings and everything in their path, but are under 3 km in diameter so damage is confined to a small area.
● **Hurricanes** – also known as tropical cyclones or typhoons, these are much larger storm systems, created by intense low pressure over warm seawater, bringing strong winds and heavy rain. They can cause floods by creating a storm surge at the coast. Hurricanes can destroy buildings over a wide area, killing large numbers of people.

Case study: Hurricane Irma, September 2017

Fact file:
● Category 5 (strongest).
● Wind speeds 185 mph.
● 134 fatalities, $65 billion damage, economic cost $300 billion.

Impacts

Impact on people	Impact on property
134 people were killed	Irma damaged or destroyed 95% of the structures on Barbuda, including its hospital, schools and both of its hotels
Many people were left without services, e.g. 362,000 customers in Puerto Rico lost water services, and more than 7.7 million homes and businesses in Florida were left without electricity	Property damage on Barbuda to the value of $300 million
The leisure and hospitality industries in the US were affected, with the loss of 111,000 jobs in September 2017	Total damage $65 billion
Florida ordered 6.5 million people to evacuate	
77,000 people were living in 450 shelters	

Exam tip

You need to know several of these impacts, from each column, including at least two place names or numbers for each. In an exam question, try to explain the impacts fully. For example, 95% of structures on Barbuda were destroyed, including the hospital and schools, meaning a lot of money was needed to replace them. 134 people were killed, which meant many families were traumatised and distressed at the loss of their loved ones.

Revision activity

Make a postcard-sized summary of your case study. Check what you need to know. Make sure you include at least two pieces of detailed information, like place names or numbers, for everything.

Now test yourself

1 How many people were killed?
2 What was the name of the one island badly affected?
3 How many people had to evacuate in Florida?
4 How strong were the winds?
5 What was the total cost of damage?

Exam practice

1 Choose one of the following, and describe two impacts it may have: drought, tornado, hurricane. [4]
2 For one case study of extreme weather, describe the impacts it had on people. [6]

Plate tectonics theory

You need to be able to:

- describe the structure of the Earth – inner and outer **core**, mantle and crust
- know that the crust is made up of a number of plates
- understand how convection currents cause plate movement
- understand the processes that occur at different plate margins
- explain what landforms are found at the different plate margins and how they are formed.

Structure of the Earth

REVISED ●

The Earth's structure can be divided into inner core, outer core, mantle and crust, as shown in Figure 1.

Layer	Characteristics
Crust	Made up of segments, a bit like the hexagons on a football, called **tectonic plates**. These float on the mantle and move about 7 cm a year. Oceanic crust (under oceans) is very dense. Continental crust is less dense.
Mantle	Semi-liquid, made of molten (melted) rock
Outer core	Semi-liquid metals such as nickel and iron
Inner core	Made of melted metal, solid due to immense pressure. Radioactive decay here generates heat

How do plates move?

REVISED ●

The heat from the radioactive decay in the inner core heats the mantle and causes convection currents, where the hottest magma rises towards the surface. At the top of the mantle, the currents spread out and drag the plates slowly apart (see plates B and C in Figure 1). When the convection currents cool and begin to sink back towards the core, they drag the plates above them closer together (see plates A and B in Figure 1).

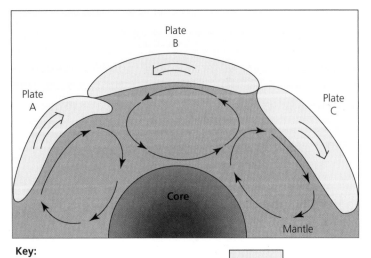

Key:

→ Convection current

⇒ Direction of plate movement

▭ Earth's crust

Figure 1 Convection currents in the mantle.

Revision activity

From the Glossary on pages 120–25 learn the definitions of tectonic plate and a convection current. Repeat them aloud to a pet, a parent or another member of your family.

Revision activity

Next time you eat an apple, cut it in half. Try to identify the core (divide it into inner and outer in your head). Make the main bit of the apple that you would eat represent the mantle. The apple skin is the crust. Try the same thing for the layers of a Mars bar – and make the chocolate move around on the caramel layer like plates on the mantle!

Types of plate margin

The edge of a plate, where it meets another, is called a plate margin or boundary. What happens at the plate margins depends on the direction of movement and the type of crust involved.

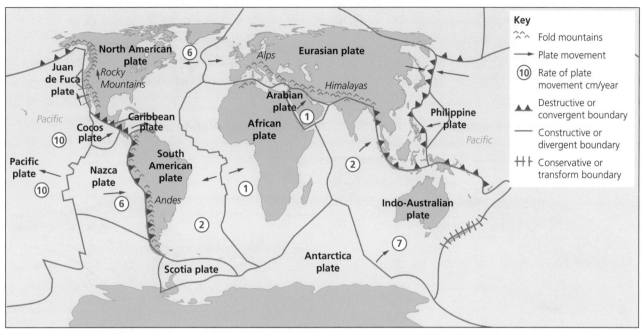

Figure 2 Tectonic plates.

Key
- ⌃⌃⌃ Fold mountains
- → Plate movement
- ⑩ Rate of plate movement cm/year
- ▲▲ Destructive or convergent boundary
- — Constructive or divergent boundary
- ┼┼ Conservative or transform boundary

What happens at different types of plate margin?

	Destructive plate margin	Constructive plate margin	Conservative plate margin	Collision zone
Where are the plates moving?	Plates move towards each other	Plates move apart	Plates slide past each other	Plates move towards each other
Where does the name come from?	Crust is destroyed	New crust is formed or constructed	Crust is conserved – neither destroyed nor added to	Two continental plates collide
What happens?	When plates meet, the oceanic plate is forced to bend and go down into the mantle beneath the other plate. This is the process of subduction and it triggers earthquakes as the plate bends. Friction and the heat of the mantle melt the descending plate in the subduction zone, forming magma. This magma rises and forms volcanoes on the continental plate	As the two plates move apart, molten rock or magma rises from the mantle to fill the gap, forming new crust	Two plates slide past one another, along a fault line. Friction between them means that they tend to stick until pressure builds up and is released in a sudden jerking movement: an earthquake	Rocks on the surface, sometimes including sedimentary rocks made on the ocean floor, fold upwards and create fold mountains. Sometimes magma starts to rise through the mountains but cools before it reaches the surface, making intrusive igneous rock underground
What features does it make?	Earthquakes, volcanoes, fold mountains and an ocean trench at the subduction zone	Earthquakes, volcanoes and mid-ocean ridge	Earthquakes along the fault line, but no volcanoes	Fold mountains Earthquake activity

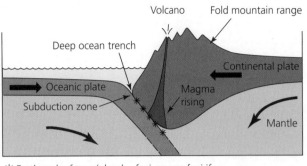

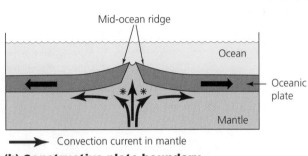

* Earthquake focus (plural = foci, so use foci if labelling a group and focus for labelling one)

(a) Destructive plate boundary.

(b) Constructive plate boundary.

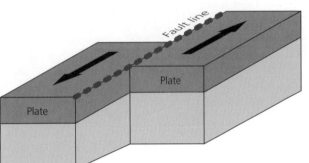

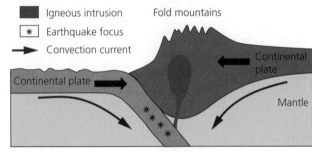

(c) Conservative plate boundary.

(d) Collision zone.

Figure 3 Four types of plate boundary.

Now test yourself   TESTED ◯

1 Draw and label a diagram showing a destructive plate boundary.
2 What causes the plates of the Earth's crust to move?
3 If plates move apart, is the plate margin constructive or destructive?
4 At what type of plate margin is an ocean trench formed?
5 At what type of plate margin is a mid-ocean ridge formed?
6 What term means the process of one plate bending under another and being forced down into the mantle?
7 What plate movement happens at a **conservative margin**?

Exam tip

Make sure you can explain what is happening in each diagram, using all the key words from page 53.

Exam practice

1 Describe the characteristics of the mantle.   [2]
2 Explain how plates move.   [4]
3 Study Figure 3a above which shows a destructive plate boundary. Describe what happens at this type of boundary.   [5]
4 Name one landform made at a constructive plate boundary and explain how it is made.   [4]

Exam practice answers and quick quizzes at **www.hoddereducation.co.uk/myrevisionnotesdownloads**

Basic rock types

You need to know

● how the basic rock types are formed and recognise their characteristics.

Formation of the basic rock types

REVISED

Figure 4 shows the three basic rock types and how they were made.

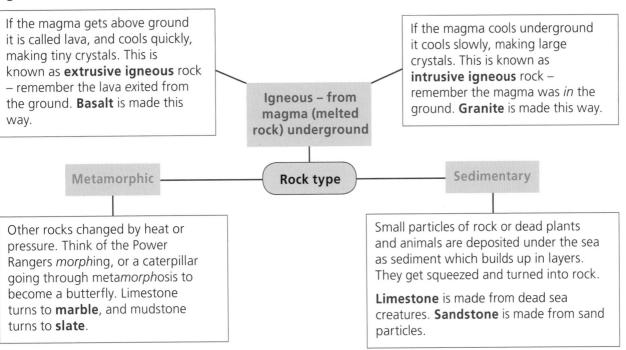

If the magma gets above ground it is called lava, and cools quickly, making tiny crystals. This is known as **extrusive igneous** rock – remember the lava *exited* from the ground. **Basalt** is made this way.

If the magma cools underground it cools slowly, making large crystals. This is known as **intrusive igneous** rock – remember the magma was *in* the ground. **Granite** is made this way.

Igneous – from magma (melted rock) underground

Metamorphic

Rock type

Sedimentary

Other rocks changed by heat or pressure. Think of the Power Rangers *morph*ing, or a caterpillar going through meta*morph*osis to become a butterfly. Limestone turns to **marble**, and mudstone turns to **slate**.

Small particles of rock or dead plants and animals are deposited under the sea as sediment which builds up in layers. They get squeezed and turned into rock.

Limestone is made from dead sea creatures. **Sandstone** is made from sand particles.

Figure 4 How rocks are made.

Characteristics of rocks

REVISED

The table below shows the characteristics of some rocks. You need to be able to recognise the rocks from the characteristics, and describe the rocks.

Now test yourself

Try to complete the last column of the table below from memory.

TESTED

Rock	Colour	Other features	Rock type: igneous, sedimentary or metamorphic?
Basalt	Dark grey/black	Glittery speckles	
Granite	Speckled grey, white, black, pink	Very hard, large crystals visible	
Limestone	Grey, white or yellow	May have fossils, fizzes when a drop of acid is added	
Sandstone	Yellow/orange	Often see grains of sand, may rub off	
Slate	Dark grey	Layers split apart easily, smooth, can be marked	
Marble	White, swirls of colour	Can be highly polished for fireplaces and floors	

55

1 Figure 5 shows how different rock types are formed. Decide which rock type belongs in each space, choosing from the words in bold in Figure 4.
2 Which category of rocks (igneous, sedimentary or metamorphic) forms from sediment that builds up in layers?
3 Which igneous rock, basalt or granite, forms from magma cooling slowly underground?
4 Name two examples of sedimentary rock.
5 Name a metamorphic rock that forms: (a) under pressure and (b) by extreme heat.

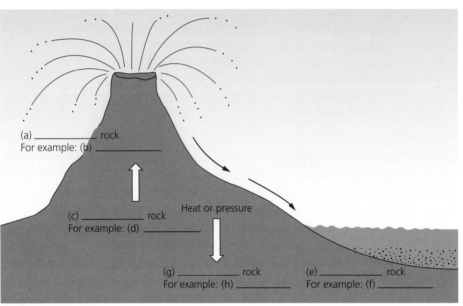

(a) _____ rock
For example: (b) _____

(c) _____ rock
For example: (d) _____

Heat or pressure

(g) _____ rock
For example: (h) _____

(e) _____ rock
For example: (f) _____

Figure 5

Revision activity

1 Search for images of each rock (basalt, granite and so on) – try to spot the main features of each one in the photo.
2 Get someone to read out the colour and other features for each rock from page 55. Try to name the rock.

Exam practice

1 For each rock, tick the correct column. [3]

Rock	Igneous	Sedimentary	Metamorphic
Granite			
Limestone			
Marble			

2 Explain how metamorphic rocks such as slate are formed. [3]
3 Read the descriptions below and give the name of the rocks. [2]

Description	Rock name
Grey, black, and white crystals	
May contain fossils, grey or white	

Managing earthquakes

You need to be able to:

● understand the causes and global distribution of earthquakes, in relation to plate boundaries
● distinguish between the focus and epicentre of an earthquake
● explain that earthquake magnitude is measured on a seismograph using the Richter scale
● understand how liquefaction and tsunamis are caused by earthquakes
● understand the causes and impacts of an earthquake using a case study from an MEDC or LEDC.

An **earthquake** is a shock, or series of shocks caused by a sudden Earth movement. The shockwaves make the ground shake.

Causes and global distribution of earthquakes

REVISED

Figure 6 shows that the distribution of earthquakes is linear (in lines or zones) and closely related to plate margins. They are mainly found:

● around the Pacific Ocean
● roughly in a north–south line in the middle of the Atlantic Ocean
● roughly in an east–west line across southern Europe, the Himalayas and South-East Asia.

Earthquakes happen at plate margins because this is where friction occurs and stresses build up as sections of crust move past each other. Stronger earthquakes happen at conservative and destructive margins.

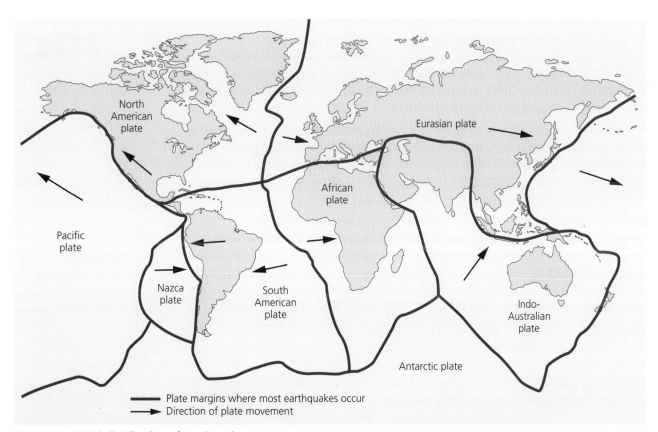

Figure 6 Global distribution of earthquakes.

Focus and epicentre

The focus is the point in the Earth's crust where the earthquake occurs and the epicentre is the point on the Earth's surface directly above the focus, where its effects are felt first.

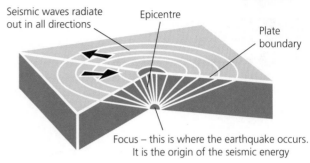

Seismic waves radiate out in all directions

Epicentre

Plate boundary

Focus – this is where the earthquake occurs. It is the origin of the seismic energy

Figure 7 Features of an earthquake.

Measuring earthquakes

An instrument called a seismograph records shock waves from an earthquake, and the magnitude of the earthquake is measured on the Richter scale. This measures the amount of shaking. Magnitude 3 can just about be felt by people who are sitting still. Magnitude 4 is ten times as strong as magnitude 3. Magnitude 5 is ten times as strong as magnitude 4, and would damage poorly constructed buildings. At magnitude 9 there is near total destruction.

Now test yourself

1 Name the point in the Earth's crust where an earthquake occurs.
2 Name the point on the Earth's surface directly above where the earthquake occurs.

Physical consequences of earthquakes

Liquefaction happens when rock or soil containing water is shaken by an earthquake. The water rises to the surface and turns the soil to liquid mud. Any buildings resting on this sink into the mud and collapse.

A tsunami is a large wave of seawater, caused by an earthquake under the sea. The shock waves from the plate movement lift the ocean, and create small waves (about 30 cm high) on the ocean surface. As they move towards the shore, and come into shallower water, they slow down due to friction and build up to become much larger – up to 30 m high.

Before a tsunami, people on the shore observe a dramatic fall in sea level, as water near the shore is pulled back into the wave. This can act as a warning.

Revision activity

Put a plate in the bottom of a sink or bowl, then fill the container with water. Quickly lift the plate up a little way and watch the waves it creates on the water's surface travel outwards.

Now test yourself

1 Name the process where water rises to the ground surface during an earthquake, causing buildings to collapse.
2 Name the type of wave that can be caused by a strong earthquake underwater.

Revision activity

To help you remember the physical consequences of earthquakes, fill a small cereal bowl with sand and add enough water to make it damp. Place three or four Lego bricks on the sand to represent a building. Set the bowl on a table and, using both hands, tap opposite sides of the bowl repeatedly and watch what happens to (1) the sand surface and (2) the Lego building. Try to explain what is happening.

Exam practice answers and quick quizzes at **www.hoddereducation.co.uk/myrevisionnotesdownloads**

Causes and impacts of an earthquake: earthquake in an MEDC or an LEDC

You need to be able to:

● know what caused the earthquake, including the names of the plates involved
● describe the short- and long-term impacts on people and the environment
● evaluate how the country prepared for and responded to the earthquake (immediate and long term).

Case study: An earthquake event in an MEDC – Tohoku

● Date: 11 March 2011.
● Magnitude: 9.1.
● Causes: The Pacific plate is subducting under the Eurasian plate, at a destructive plate margin, moving at 8–9 cm a year. Pressure was bending the Eurasian plate down, and eventually it broke along the boundary, 15 miles under the surface. This created shockwaves, which made the ground shake for 5 minutes, and lifted the sea floor up 7 metres, 80 miles off the coast of Sendai, causing a large tsunami.

Impacts

Impacts	Short term	Long term
People	● 332,395 buildings, 2126 roads, 56 bridges and 26 railways were destroyed or damaged. 300 hospitals were damaged and 11 were totally destroyed. ● 500,000 people were left homeless, losing most of their possessions and having to uproot their lives. ● 15,894 people died, 6152 people were injured, 130,927 were displaced and 2562 people remain missing. This left many people in mourning.	● The damage caused cost $360 bn to repair. Japan had to go into debt and people paid high taxes for a long period. ● The damage caused by the earthquake resulted in the meltdown of seven reactors.
Environment	● The Fukushima nuclear power plant overheated and released significant radiation into the air and water. Water sampled near the plant's seawater discharge point contained 4385 times more radiation than the maximum safety levels.	● A 250-mile stretch of coastline dropped by 0.6 m, allowing the tsunami to travel further inland. ● The earth's axis moved by between 10 and 25 cm, shortening the day by 1.8 microseconds.

Preparation for and response to the earthquake

	Measures	Evaluation
Preparation	● Japan has had a tsunami warning system since 1952. There is also a network of seismographs across the country to monitor and report any earth movements. ● Billions of pounds have been spent making buildings more resistant to earthquakes. This involves using types of glass that does not shatter, weights in buildings to counter the sway or huge shock absorbers in the foundations.	+ The warning system operated as it should, interrupting TV broadcasts, giving a few seconds warning of shaking, and about 20 minutes for the tsunami. − However, people did not realise the scale of this earthquake, so some did not take action. Only 58% of people moved to high ground before the tsunami – some thought the tsunami walls would protect them. + Many buildings survived the earthquake itself, − but were destroyed by the tsunami.
Immediate and long-term strategies	● The army helped to build many temporary shelters very quickly and provided food, water and medicine. ● The rebuilding of the worst affected areas began almost immediately. The government set up a Reconstruction Design Council which had a budget of over 23 trillion Yen to rebuild houses.	+ 163 countries offered help in the aftermath of the earthquake. − Some shelters initially were poorly equipped but this was improved quickly. + By 2018, house building was 94% complete and 99% of roads were rebuilt. − However, some people lived for years in temporary prefabricated housing.

Revision activity

Make a postcard-sized summary of your case study. Check what you need to know. Make sure you include at least two pieces of detailed information, like place names or numbers, for everything.

Exam practice

1 Explain how earthquakes can happen at conservative plate margins. [3]

2 Describe the pattern of earthquakes around the world. [3]

3 Explain how an earthquake can cause a tsunami. [4]

4 For a case study of an earthquake, name the plates involved and explain the causes of the earthquake. [4]

5 For an earthquake you have studied, discuss the long-term impacts it had on people and the environment. [5]

6 For an earthquake you have studied, evaluate how the country prepared for the earthquake. [5]

7 For an earthquake you have studied, evaluate the long-term strategies implemented after the event. [5]

Now test yourself   TESTED

1 Name the plates involved in the 2011 earthquake in Tohoku, Japan.

2 What was the magnitude of this earthquake?

3 What was the cause of thousands of deaths from this earthquake?

4 Approximately how many people were left homeless by it?

5 How effective was the preparation for an earthquake? Give some evidence.

6 Evaluate the effectiveness of the response to the earthquake.

Revision activity

Make a copy of all the bullet points in the previous two tables on page 59. Jumble them up and then practise putting them under the correct heading: impacts or management response, short term or long term, people or environment.

Volcanoes – characteristics and consequences

Summary

- describe the characteristics of shield, composite and supervolcanoes
- discuss the potential global impacts of a supervolcano eruption on people and the environment.

Volcanoes are mountains, often cone-shaped, formed by surface eruptions of magma from inside the Earth.

Types of volcano

REVISED

There are three main shapes of volcano:

- Shield volcano – low and wide, with gentle slopes. This is because they are made from very runny lava, at constructive plate margins, which can travel a long way before it cools and becomes solid rock. Think about spilling a jug of runny custard over the table – it will run a long way. Eventually layers build up to create a low, wide volcano.
- Composite volcano – higher and narrower, classic mountain shape. This is because they are made from thicker lava, at destructive plate margins, which becomes solid before it can travel too far. Think of a jug of very thick custard – it doesn't run far. These volcanoes usually release ash as well as lava when they erupt, so they will be made of alternating layers of each.
- Supervolcano – these are much bigger, and an eruption releases about 1000 times the amount in a normal volcanic eruption. They happen rarely – eruptions may be hundreds of thousands of years apart. A huge magma chamber builds up, creating a massive eruption where magma, ash and gases escape through cracks or fissures in the ground. The magma chamber is then emptier, so the land above it collapses, leaving a big depression, called a caldera. Yellowstone, in north-western USA, is a supervolcano which has erupted three times in the last million years. The next eruption will have huge impacts on the world.

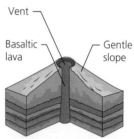

Figure 8 Shield volcano

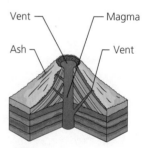

Figure 9 Composite volcano

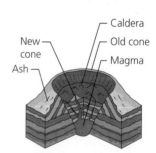

Figure 10 Supervolcano

Exam tip

Make sure you can describe each of the three main shapes of volcano, and compare them.

Now test yourself

TESTED

1. Which type of volcano is (a) high, (b) wide, (c) most destructive?
2. Give two differences between shield and composite volcanoes.
3. Give two characteristics of a supervolcano which are unique.
4. Which types of volcano include ash?
5. Which type of volcano is made of runny lava?

Consequences of volcanoes

Supervolcanoes are highly explosive, and give out at least 1000 km³ of material when they erupt. This means an eruption would have huge impacts on people and the environment, and some of these impacts would be global. If Yellowstone erupted, it is likely to affect a very large area, with around 10 cm of ash falling in an area 800 km wide.

Case study: Yellowstone

Impact on people	Impact on the environment
Everyone within 1000 km killed – inhaling ash, creating concrete-like substance in their lungs, could kill approximately 90,000 people	Global climate change for at least six to ten years, reducing average temperatures by 10°C due to sulphuric gas aerosols in the atmosphere
Many buildings destroyed by only 30 cm dry ash. This could be as far away as Los Angeles and Chicago	All mammals in Yellowstone Park likely to die, including bison and wolves, disrupting local ecosystem for many years
Vehicle filters clogged up – air and road travel severely disrupted	Monsoon rains in Asia may fail, creating drought conditions
Water supplies undrinkable	
Usual global rainfall and temperature patterns upset, which could cause harvests to fail and famines around the world	

Now test yourself

TESTED

If Yellowstone erupted:

1 How many people could be killed by inhaling ash?
2 Name two types of animal in Yellowstone Park that would be killed.
3 What effect would it have on global temperatures?
4 Name an area where the climate would change significantly.
5 What might happen to buildings? Where? How?

Exam practice

1 Read the descriptions and match them up with the correct type of volcano. [3]

Description	Type
Wide and shallow, made of layers of lava	Composite volcano
Very large caldera made from a large eruption in the past	Shield volcano
Tall, made of layers of ash and lava	Supervolcano

2 Explain how a supervolcano is formed. [4]
3 For a supervolcano that you have studied, discuss the potential global impact of an eruption on the environment. [5]

Revision activity

1 Make a postcard-sized summary of this case study. Make sure you have at least three impacts for people and three for environment, with at least two figures or place names for each one.

2 Draw a volcano in the middle of a sheet of paper. Take two colours, one for people and one for environment, and write labels to show the impacts Yellowstone could have it if erupted. Make sure you include facts and figures.

Exam tip

Make sure you use the word global in your answer to question 3.

Theme A Population and Migration

Population growth, change and structure

> **You need to be able to:**
>
> - define crude birth rate, crude death rate and natural change (increase and decrease)
> - show detailed knowledge and understanding of the demographic transition model
> - understand what is meant by population structure
> - interpret population pyramids, identifying youth-dependent and aged-dependent populations
> - compare and contrast population pyramids for an MEDC and an LEDC
> - assess economic and social implications of aged and youth dependency.

Definitions of crude birth rate, crude death rate and natural change

REVISED

Crude birth rate is:
- the number of live births
- per 1000 people
- per year.

Crude death rate is:
- the number of deaths
- per 1000 people
- per year.

The natural change is the difference between the birth rate and the death rate.

When the birth rate is more than the death rate, that means more people are being born than are dying in any year. This means the population will increase. The amount it increases by is called the natural increase. To work it out, take the death rate away from the birth rate.

Example:
- birth rate = 30
- death rate = 20
- natural increase = 30 – 20 = 10.

The natural increase is 10 per 1000, or 1% growth rate (obtained by dividing by 10).

When the death rate is more than the birth rate, more people die than are born in any year. The population will decrease. This is called the natural decrease. The amount it decreases by can be worked out in the same way as above.

Example:
- birth rate = 20
- death rate = 25
- natural decrease = 20 – 25 = –5.

The natural decrease is –5 per 1000, or –0.5% growth rate.

> **Exam tips**
>
> You need to remember all three parts of these definitions.
>
> 'Crude' just means basic, without any extra complicated measurements.

Demographic transition model

As countries become more developed, their birth and death rates change, resulting in a huge population change – a big increase in the total population. We can see this in the demographic transition model, which is a way of summarising what has happened in most countries.

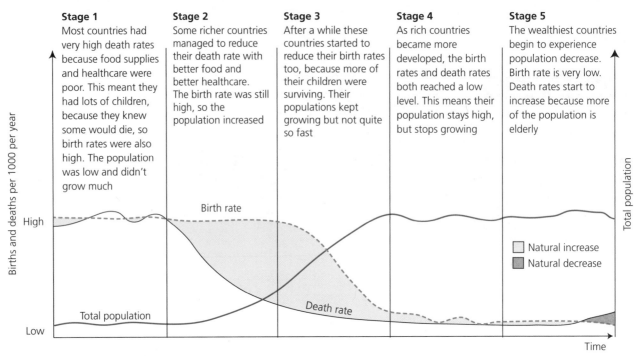

Stage 1
Most countries had very high death rates because food supplies and healthcare were poor. This meant they had lots of children, because they knew some would die, so birth rates were also high. The population was low and didn't grow much

Stage 2
Some richer countries managed to reduce their death rate with better food and better healthcare. The birth rate was still high, so the population increased

Stage 3
After a while these countries started to reduce their birth rates too, because more of their children were surviving. Their populations kept growing but not quite so fast

Stage 4
As rich countries became more developed, the birth rates and death rates both reached a low level. This means their population stays high, but stops growing

Stage 5
The wealthiest countries begin to experience population decrease. Birth rate is very low. Death rates start to increase because more of the population is elderly

Births and deaths per 1000 per year

Total population

Birth rate

Death rate

Total population

High

Low

Time

Natural increase
Natural decrease

Figure 1 Demographic transition model.

The population of the whole world has been affected by the changes shown in Figure 1.

Now test yourself

1 Why were death rates high in Stage 1?
2 Why did death rates fall in Stages 2 and 3?
3 Why were birth rates high in Stage 1?
4 Why did birth rates fall in Stage 3?
5 Which stages have the fastest population growth?
6 Why does death rate increase in Stage 5?

In more developed countries like the UK and USA, the populations have already increased and now stay more or less the same.

Some less developed countries like India and Mexico are now in the middle stages shown in Figure 1, and their populations are growing quickly.

Exam tip

Don't panic – the name of this model makes it sound more complicated than it really is! 'Demographic' just means to do with population. 'Transition' means a big change that takes place over time, and 'model' means it's a simplified version of the real world. So it's a simple version of the change that takes place over time in the population of a country.

Revision activity

Practise drawing this model using three different colours for birth rate (BR), death rate (DR) and total population. Make sure you can draw the lines perfectly, with the stages marked on. Add labels to explain what is happening in each stage.

This means the world population is going to continue to grow in the next few years, as shown in Figure 2.

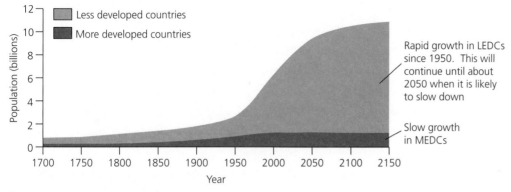

Figure 2 Graph of world population growth.

Now test yourself   TESTED ⬤

Decide which statements are true and which are false:

1 The population of the world is decreasing.
2 The population of the world is increasing.
3 World population grew slowly until 1950.
4 World population grew rapidly from 1700.
5 World population grew rapidly from 1950.
6 The population in 2000 was growing fastest in MEDCs.
7 The population in LEDCs is growing much faster than in MEDCs.
8 LEDCs are mostly in the middle stages of the demographic transition model.
9 LEDCs are mostly in the last stage of demographic transition model.

Population structure   REVISED ⬤

Population structure is the way the population is divided between male and female, and how it is divided between the different age groups.

It can be shown clearly on a population pyramid such as Figure 3. The shape of a population pyramid can tell us a lot about the population.

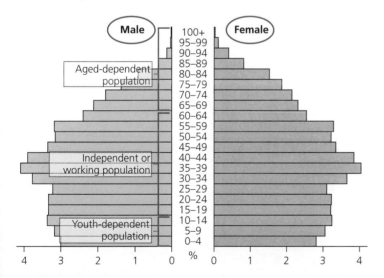

Figure 3 An example age–sex pyramid.

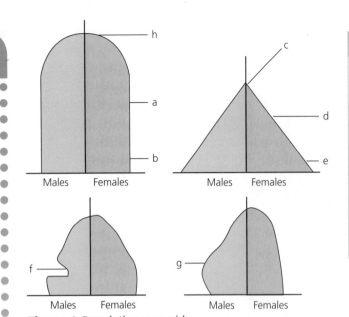

Match up the following labels to the appropriate letters on the population pyramids in Figure 4.

1 Wide base shows high birth rate.
2 Narrow base shows low birth rate.
3 Straight pyramid shows low death rate.
4 Triangular pyramid shows high death rate.
5 Tall pyramid shows high life expectancy.
6 Short pyramid shows low life expectancy.
7 Missing young adult males show out-migration.
8 Extra young adult males show in-migration.

Figure 4 Population pyramids.

Dependency can be seen clearly on population pyramids. There are two groups of people who are dependent on the people who are of working age to support them:

● aged dependent: age 65+   ● youth dependent: age 0–14.

We can calculate a dependency ratio to show the percentage of the population dependent on the rest:

$$\text{Dependency ratio} = \frac{\text{Youth dependent} + \text{Aged dependent}}{\text{Working population}} \times 100$$

Case study: Population structure of an MEDC – aged-dependent population

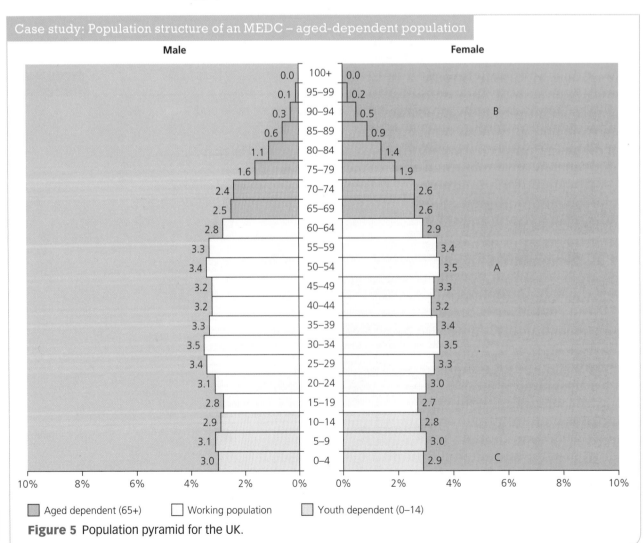

Figure 5 Population pyramid for the UK.

Exam practice answers and quick quizzes at **www.hoddereducation.co.uk/myrevisionnotesdownloads**

Now test yourself
TESTED ⚪

Match up the following labels to the letters A–C on the pyramid in Figure 5.

1 Tall pyramid shows large percentage of elderly people (19% in 2019). This is the aged-dependent population, who generally do not earn money and need to be supported by the working population.

2 Base of pyramid is getting narrower. This shows the birth rate is falling.

3 Wide section in the middle shows a large population currently of working age who will add to elderly population over the next 30 years.

Case study: Population structure of an LEDC – youth-dependent population

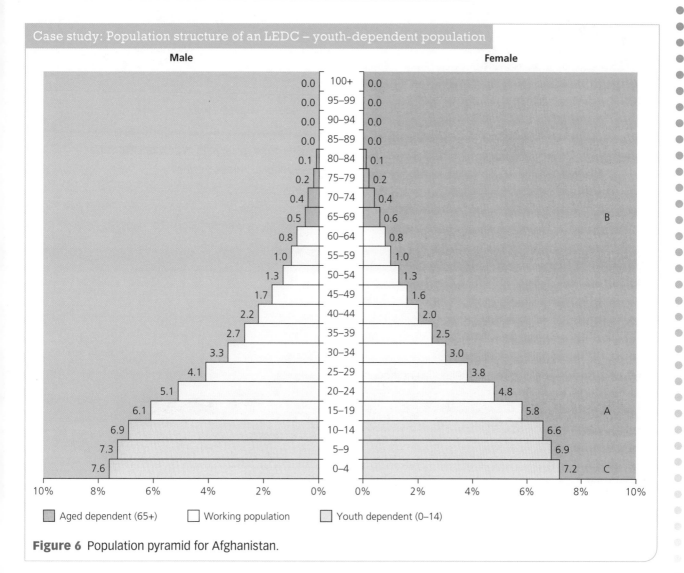

Figure 6 Population pyramid for Afghanistan.

Now test yourself
TESTED ⚪

Match up the following labels with the letters A–C on the pyramid in Figure 6.

1 Base of pyramid is getting wider. This shows the birth rate is high. This is the youth-dependent population, who do not earn money and need to be supported by the working population – 43% in 2019.

2 Lots of people of childbearing age means the birth rate will stay high.

3 Pyramid is relatively short, which means death rate is still high.

Revision activity

Make a postcard-sized summary of each case study. Sketch the shape of the pyramid, and add three labels to each one, giving the main points.

Exam tip

If you are asked to compare, you need to use comparison words – like bigger, taller, shorter, narrower, wider. For example, a pyramid for an LEDC has a wider base than one for an MEDC, and is shorter. LEDC pyramids get narrow more quickly, and are narrower at the top than MEDC pyramids.

Exam tip

In an exam you can sketch the shape of the pyramids if it will help you answer a question.

67

Implications of aged and youth dependency

Both aged dependency and youth dependency can create problems for the country to deal with. The government gets money by taxing people on what they earn. This means people of working age are needed to provide tax money to pay for services like healthcare, education and so on. This means that if there are lots of dependent people and not so many of working age, there are economic implications (to do with money). There are also social implications (to do with the way people behave).

Revision activity

Copy and complete the tables below by putting the impacts from the box into the correct part of the tables. Make sure you can explain your decisions. Some may fit into more than one part of the table!

Impacts:

1 Adults giving up careers to care for elderly relatives
2 Elderly can provide wise advice
3 Expensive healthcare for the elderly
4 Lack of school buildings and facilities
5 Lack of teachers
6 Large numbers of infant vaccinations needed
7 Lots of young adults entering the labour market

8 Meals on wheels and home helps
9 Pensions
10 Relatives may be able to provide childcare
11 Residential homes needed
12 Strain on carers
13 Strain on primary schools – some operate 2 half-day sessions for different groups of pupils

	Costs	Benefits
Social		
Economic		

Figure 7 Socio-economic implications of aged dependency in MEDCs.

	Costs	Benefits
Social		
Economic		

Figure 8 Socio-economic implications of youth dependency in LEDCs.

Now test yourself

TESTED

1 What age groups are defined as: (a) youth dependent and (b) aged dependent?
2 A wide-based triangular pyramid shows: (a) youth dependency or (b) aged dependency?
3 What type of dependent population results in a shortage of school buildings and teachers?
4 What type of dependent population results in increased need for pensions?
5 Suggest one social benefit of having an aged-dependent population.

Exam practice

1 State the meaning of the term 'crude death rate'. [2]

2 Complete the table showing how population changes due to birth rate and death rate. [3]

Country	Birth rate/1000	Death rate/1000	Natural change/1000
A	25	17	
B		11	3
C	13		−2

3 Study the demographic transition model diagram on page 64. Explain why population may start to decline in Stage 5. [4]

4 Study the population pyramid below and state whether this is likely to be from an LEDC or an MEDC, and give evidence for this. [4]

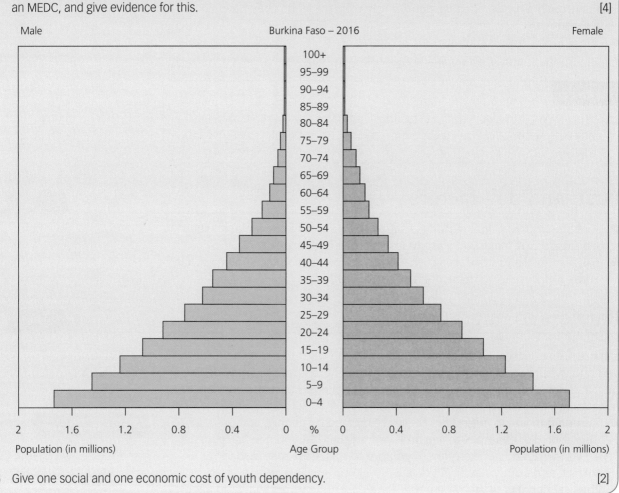

Male   Burkina Faso – 2016   Female

Age groups (top to bottom): 100+, 95–99, 90–94, 85–89, 80–84, 75–79, 70–74, 65–69, 60–64, 55–59, 50–54, 45–49, 40–44, 35–39, 30–34, 25–29, 20–24, 15–19, 10–14, 5–9, 0–4

Male scale: 2, 1.6, 1.2, 0.8, 0.4, 0 — Population (in millions)
Female scale: 0, 0.4, 0.8, 1.2, 1.6, 2 — Population (in millions)
% Age Group

5 Give one social and one economic cost of youth dependency. [2]

Causes and impacts of migration

Migration

REVISED

Migration means permanent movement of people from one place to another. This is usually a big decision, so people only migrate when they have good reasons to.

- immigration means moving into a country.
- emigration means moving out of a country.

Exam tip

Make sure you get these the right way round! **I**mmigration – moving **I**n. **E**migration – moving out – like **e**xit. Try to get the spelling – two 'm's for immigration, only one for emigration.

Push and pull factors

REVISED

- Push factors are negative things that push you away from a place – they are things that make you want to leave it, like high crime rates.
- Pull factors are positive things that pull you towards a place – things that make you want to go there, such as better-paid jobs.

Barriers to migration

REVISED

If there are enough push factors making someone want to leave their home, and pull factors making them want to go somewhere else, they may decide to migrate. However, sometimes other things stop them. These are barriers to migration:

- **Human barriers to migration:** these are created by people, and prevent migration. For example, to live and work in the USA you need a special kind of visa, and not everyone is entitled to this. Cost may also act as a barrier, and family ties.
- **Physical barriers to migration:** these are natural features that stop people moving from one place to another. The barrier may be mountains, the sea, a river, a desert. This is sometimes called topography, meaning the arrangement of natural features.

Revision activity

1. Make two revision cards – one for push factors, one for pull factors. Include a definition and at least three examples of each.

2. Imagine you are deciding whether to migrate. Make up the story – make it funny or silly to be more memorable! Where do you live? What are the push factors? Where do you want to go? What are the pull factors? What physical and human barriers are there to migration?

Now test yourself

TESTED

Colour-code the list below. Use red for push factors, green for pull factors, blue for human barriers to migration, black for physical barriers to migration:

Few jobs
Poor education opportunities
Poor healthcare
Good jobs available
Better education
Safer environment

Drought
Mediterranean Sea
Mountains to cross
Visa requirement
No entitlement to work in
    destination country

Economic migrants and refugees

- An economic migrant is someone who moves because they want to improve their chance of getting employment and earning money.
- A refugee is someone who is fleeing wars or persecution and has been granted protection by their destination country – the country they have moved to.

Challenges faced by refugees and their destination country

Case study: Greece

Basic facts:
- Refugees travel via Turkey from Syria, as well as Afghanistan and Iraq, to get to Europe, fleeing from war zones.
- Other refugees come from African countries such as Eritrea and Somalia.
- Most arrive by boat after a dangerous journey (60,000 in 2019).
- 240,000 arrived in January–August 2016. About a quarter of these were children. Although numbers are now smaller, this is continuing – 75,000 arrived in 2019.
- Many migrants became stuck in Greece as other European countries prevented them travelling further.

Challenges faced by refugees	Challenges faced by Greece
Dangerous journey – paying people-smugglers to take them across the Mediterranean, often in boats which are overcrowded. In 2016, 441 died on the way	Poor economy – 18% unemployment in 2020, and loss of tourism revenue during the pandemic, means they have limited resources to provide for refugees
Living in overcrowded camps, such as 20,000 people living in the Moria camp, designed for 3000, with limited food and shelter; up to 30 people living in a shelter made from a shipping container. During the coronavirus pandemic there has been limited opportunity for distancing and hygiene. Much of the camp was destroyed in a fire in 2020, leaving 13,000 people with no shelter	Limited support from other European countries, preventing migrants from moving further north. This results in large numbers of refugees being stuck on Greek islands, such as 20,000 in the Moria camp on the island of Lesvos
Limited education for children and no chance to put down roots and find jobs	Anti-immigrant sentiment rising in Greek population as people fear that the refugees will take their jobs and resources – neo-Nazi Golden Dawn Party supporters and 16 other right-wing groups have attacked migrants
Threatened with being sent back – there are allegations that Greece has been abandoning asylum seekers on inflatable rafts	Large numbers of refugees – 185,000 in 2019

Now test yourself

1 How many migrants arrived in Greece between January and August 2016?
2 Name three countries the migrants come from.
3 Name a refugee camp in Greece.
4 What is the unemployment rate in Greece?
5 How many refugees were in Greece in 2019?
6 Give two challenges for the migrants.
7 Name the neo-Nazi party in Greece whose supporters have attacked the migrants.

Revision activity

Make a postcard-sized summary of this case study. Make sure you include the challenges for the refugees, the challenges for Greece, and at least three pieces of case study detail – place names or figures.

Exam practice

1 State the meaning of 'immigration'. [2]

2 What is the difference between an economic migrant and a refugee? [3]

3 Read the account below and then answer the questions.

Carlos was an unemployed builder who decided to move to the USA to look for work and to stay with distant relatives who already lived there. His plan was to earn money to send to his family and parents. He left El Salvador, travelling the 3000 km to Austin, Texas, by road. He never reached the USA: border guards intercepted him and returned him to El Salvador.

(a) Give one pull factor that made Carlos migrate. [1]

(b) Give one barrier to migration that Carlos encountered. [1]

4 With reference to your case study, discuss the challenges faced by the destination country. [6]

Urban land use in MEDC cities

Land use zones

Land use means what the land is used for, for example, shops, industry or housing. Land use zones are areas where the land is used mainly in one way. The following diagram shows how the main land use zones are commonly arranged in an MEDC city.

1 CBD
2 Inner city
3 Suburbs
4 Rural–urban fringe

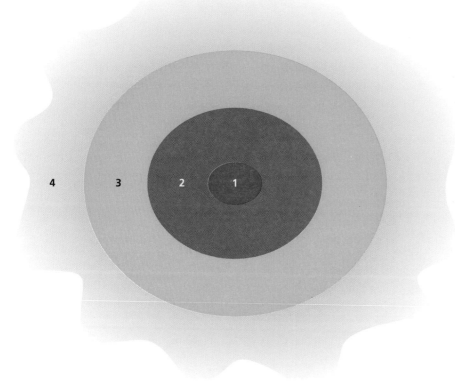

Figure 1 Land use zones in an MEDC city.

Land use zone	Characteristics
Central business district (CBD)	• Shops, offices, entertainment • Easy access by bus, train or car • Land is expensive because space is limited, so people build skyscrapers
Inner city	• Originally factories and small terraced houses in tightly-packed streets • Some buildings are old, dilapidated and boarded up, awaiting redevelopment • Some former houses are converted into small offices • Large houses are divided up to make smaller flats, often for students • Some terraced housing areas have been demolished and replaced with high-rise blocks of flats • Some areas are rebuilt or converted into large expensive apartments – this is called **gentrification**
Suburbs	• Mainly housing, with some light industry and shopping • Often detached or semi-detached houses, in streets including cul-de-sacs and curved streets
Rural–urban fringe	• A zone of mixed urban and rural land uses • Space available for a golf course, waste recycling centre, hospital or airport • Sometimes urban sprawl means that the town or city can gradually take over the countryside

Now test yourself

TESTED

1 Which zone would be most likely to have:
   (a) the most shops?
   (b) the highest buildings?
   (c) mostly houses?
   (d) old buildings boarded up?
   (e) a golf course?
2 Name the zone at the centre of a city.
3 Name the zone at the edge of a city.

Revision activity

Create your own land use zone diagram of a city. On a large sheet of paper, draw a diagram with four concentric circles, like Figure 1. Use an internet search to find pictures of things you might find in each land use zone. Print, cut out and stick – or create a diagram on the computer and print it out. Add labels for the zones, and some descriptive words.

Interpretation of maps and aerial photographs

A settlement is a place where people live – a village, town or city. We can spot the land use zones of cities on aerial photographs and on Ordnance Survey (OS) maps. We can also find evidence of the settlement function – the job it does. For example, some settlements are ports, or tourist towns, or have a lot of industry, and some areas are residential.

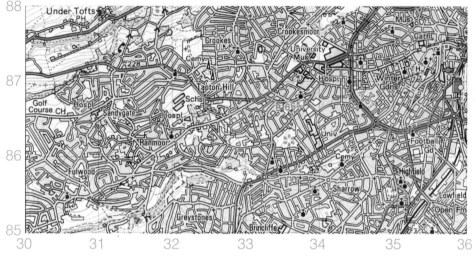

Figure 2 Ordnance Survey map of Sheffield (scale 1:50,000).

> **Exam tip**
>
> Make sure you can use four-figure grid references (see page 10) – an exam question might ask you to identify the land use zone in a particular square. Remember – the numbers are the coordinates of the bottom left-hand corner of the square.

Revision activity

1. In the map in Figure 2, find the following:
   - Main roads and public buildings – CBD.
   - Parallel streets close together – inner-city terraced housing.
   - Bendy roads, with dead ends – suburbs.

2. Using the map on the inside back cover, look at Cushendall and see if you can identify any functions of the settlement.

3. Match the photos here to the following descriptions:
   a) Shadows tell us buildings are tall – CBD.
   b) Narrow parallel streets, little green for gardens – inner city.
   c) Varied street shapes, including dead ends and gardens – suburbs.

1   Give three possible functions of a town.
2   What would shadows on an aerial photo tell you about a building?
3   Which zone would have curved streets?
4   On a map, which zone would have lots of red roads meeting together?

		On an OS map	On an aerial photo
Function	Tourism	Look for tourist facilities – usually in blue, such as: • viewpoint • car park　P • museum　Mus • castle　**Castle** • information centre　**i**	Beaches will show up yellow or grey. Car parks may be visible
	Port	Port facilities – labelled harbour, or showing ferry routes coming in and out of them, regular rectangular shapes along the coast indicating docks or a harbour wall Ferry P　Ferry V	Regular rectangular shapes at the coast. Ships may be visible
	Industry	Large buildings, often grey, sometimes labelled 'works'	Large grey buildings
	Residential	Built-up area with streets. Inner-city area will have lots of parallel streets. Suburbs will have dead ends, curved road pattern	Houses and road patterns, possibly gardens
Land use zones	CBD	Lots of main roads (red) meeting. Often a train station – black railway line and red circle or rectangle for station. Rectangular street pattern A 493	High buildings found in many CBDs often have shadows. Buildings may cover large areas, and key buildings such as a town hall may have distinctive features such as domed roof or formal gardens
	Inner city	Parallel streets in a built-up area, possibly with industry, such as the streets in Figure 2, grid square 3287	Parallel rows of terraced houses with no gaps in between them and small gardens
	Suburbs	Street patterns with curved streets and dead ends, such as the roads in Figure 2, grid square 3185	Semi-detached and detached houses, with green around them
	Rural–urban fringe	Edge of built-up area. Blue flags for golf courses. Hospitals and airports often marked	Edge of housing area, fields. Sand bunkers on golf courses may show up. Hospitals show up as large buildings with car parks. Airport runways may be visible

Exam practice

1 Give two things you would expect to find in the CBD of a city. [2]

2 Study the photograph below and answer the questions that follow.

(a) Give two pieces of evidence from the photograph that A is the CBD. [2]

(b) What land use zone is found at B? [1]

3 Using the OS map on the inside back cover, give two functions of the town of Cushendall. [2]

Issues facing inner-city areas in MEDCs

You need to be able to:

● explain issues associated with inner cities in MEDCs – housing, traffic and the cultural mix.

Housing

Poor-quality housing

Terraced houses were built originally for factory workers. They were small and close together so have limited space for children to play

Older buildings are not very efficient – their roofs are not well insulated and they do not have cavity walls so may be damp and draughty

Many houses were run down and became slums, which were demolished in the 1960s. These were built back to back, with no back door, no garden and a shared toilet. Birmingham had a lot of these houses

The strong communities which built up in these areas have declined in many areas as people move out. Unemployment is often high as the factories have closed

When these houses were first built there were no cars. Now streets are busy, narrow and dangerous for children

Some houses were well built but needed to be renovated, with added kitchens and bathrooms. This includes many areas in Belfast, such as the lower Newtownards Road

Gentrification

Inner-city areas have declined as factories closed, houses were knocked down to remove slums and people moved away.

Urban regeneration has been attempted in many cities. This means that governments spend money to improve the area, and attract new businesses, creating jobs, in the hope of bringing new life to the area.

As part of urban regeneration, many older buildings, such as factories and houses, are bought by developers and renovated for people on high incomes, for example, creating expensive apartments. This is called gentrification.

Advantages	Disadvantages
The area looks better, which is more likely to attract businesses into the area, creating jobs	Original communities are pushed out as they cannot afford the new housing. This is called gentrification
New shops, cafés and other services are more likely to locate in regenerated areas	New cafés are useful to the richer residents but not so useful to the original residents
Maintains old architecture	The jobs created in the area often do not suit the skills of the older original residents

Now test yourself

1 List three problems with the old houses in inner-city areas.
2 What is meant by 'gentrification'?
3 Give two problems with gentrification.

Exam practice answers and quick quizzes at **www.hoddereducation.co.uk/myrevisionnotesdownloads**

Traffic

Congestion

Traffic in inner-city areas is often congested because:

- streets are narrow, as they were built when most people walked
- inner-city residents often have no garages or driveways so park their cars on the road
- people driving into the CBD may try to park in inner-city streets to avoid parking charges.

This has impacts on air quality and journey times. There has been some evidence of reduced congestion and pollution as more people worked from home during the coronavirus pandemic. It is too early to tell whether these changes are likely to be permanent or temporary.

Diesel engines produce tiny particles called particulates. These are absorbed by the lungs and are responsible for deaths

Nitrogen dioxide, mainly from car pollution, affects people's lungs, causing 28,000 premature deaths in the UK each year

In London, congestion adds about 1.4 minutes for every kilometre travelled

Cities have tried to improve air quality – Mexico City and others are banning diesel vehicles. London has an Ultra Low Emission Zone, charging vehicles which cause pollution, to discourage them from driving through the zone

Congestion in London costs about £9.5 billion a year (2017) as business vehicles are delayed and working time is spent in traffic jams

Increased journey times create more pollution and reduce quality of life as more time is spent commuting. The chances of accidents increase

Now test yourself

TESTED

1 Give three reasons why inner cities experience traffic congestion.
2 How many deaths in the UK each year are caused by nitrogen dioxide?
3 Name a city which is banning diesel vehicles.
4 How much does congestion cost each year in London?
5 How much time is added to each kilometre travelled in London due to congestion?

Exam tip

Make sure you know a few details, like figures or place names, as these will make your exam answers more precise and get you extra marks!

Public transport

Inner-city residents are often dependent on public transport, such as buses, trams and trains. There are two main issues here: cost and efficiency.

Cost

Public transport can be expensive, taking a significant proportion of someone's wages to travel to and from work. In London the average commuter usually spends nearly 20% of their salary on travelling to work

Cheaper housing is found at the edge of the city – but this may increase the cost of travel to work in the CBD

Efficiency

Using public transport can be inefficient as time may be wasted, particularly waiting for connections, or having to travel into the city centre first and then out

Choices may be limited – London has only been running tube trains at night from October 2016 – before that, night buses served limited areas and were slow

Parking

Inner-city areas have two main issues related to parking: cost and availability.

Cost	Availability
Inner-city parking may add up to be very expensive: £1.20 an hour works out at over £10 for a working day	Residents don't have driveways so have to park on the street. This may mean parking a long way from their house as there are limited spaces
Residents may have to pay for resident-only parking	Drivers may circle round looking for a parking space, creating more congestion and more pollution. In Los Angeles a smart app can now tell drivers where they can find a parking space
Fees for parking illegally may be very high – some areas will tow a vehicle away and demand a fee of £160 to release it	Residents may struggle to get a space as commuters may park their cars in the inner city. Some areas have introduced resident-only permit systems to try to combat this

Now test yourself

TESTED ◯

1 How much could you pay if your car gets towed away?
2 Name a city with a smart parking app.
3 How have cities tried to ensure residents can park in inner-city areas?

Exam practice

Study the graph below, which shows scheduled (planned) journey time and actual journey times by bus.

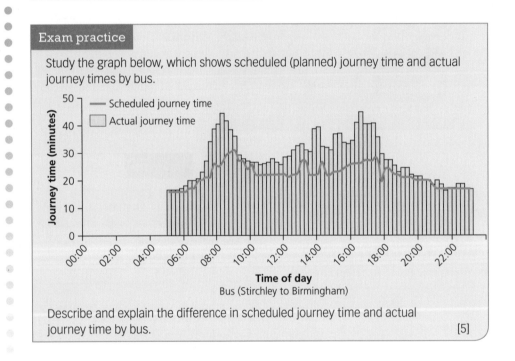

Bus (Stirchley to Birmingham)

Describe and explain the difference in scheduled journey time and actual journey time by bus.

[5]

The cultural mix

Immigrants often move into the inner-city areas of cities. This is because there is often cheap housing there, and because immigrants grouping together allows people to set up shops, places of worship, community centres and so on which will meet their needs.

This means there are often large communities of people from different places, with different languages, cultures and religions, in the inner-city areas. As well as bringing variety and cultural richness to an area, this can create issues, including ethnic tensions, religious tensions and language barriers.

Ethnic tensions

Different ethnic groups may experience tension or conflict: between two minority groups or between one immigrant group and the host population. This could be because they do not understand each other, or because the host population is worried the newcomers will take their jobs. For example, in Belfast, there were 589 recorded racist crimes from July 2019 to June 2020. Of these, 355 involved violence against a person.

Religious tensions

Sometimes, people's loyalty to their religion can be particularly strong, creating tension with other religious groups. There were 8336 religious hate crimes in England and Wales in 2017–18, with over half directed at Muslims.

Language barriers

A mix of cultures also means a mix of languages. About 300 languages are spoken in London, some by large numbers – such as Bengali and Urdu. Many individuals, especially recent immigrants or older people, may have trouble learning English, so find it difficult to access health services, employment or benefits.

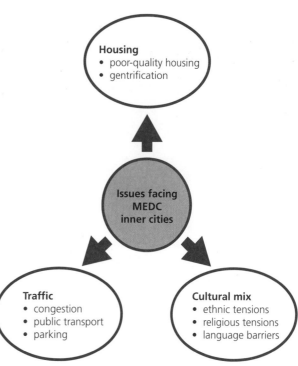

Figure 3 Issues facing MEDC inner cities.

Exam practice

1 Inner cities in MEDCs have many difficult issues to deal with. One of these is traffic. Give three reasons why traffic in inner cities can be a problem. [3]

2 Give one advantage and one disadvantage of gentrification. [2]

3 Study the photo below and answer the question. How can ethnic tensions create problems in inner cities? [3]

Revision activity

On a large piece of paper, create a spider diagram to show all the issues facing inner-city areas in MEDCs. Include figures or place names where possible. Use the outline in Figure 3 to start you off.

Urbanisation in MEDCs and LEDCs

You need to be able to:

● evaluate one MEDC urban planning scheme that aims to regenerate the inner city, using case study detail
● describe and explain the location, rapid growth and characteristics of shanty towns using a case study.

Case study: MEDC urban planning scheme – Titanic Quarter, Belfast

Titanic Quarter is a 75-hectare site in Belfast's inner city, within 10 minutes' walk of the CBD, on the east bank of the River Lagan. This is where the *Titanic* was built in the world-famous Harland & Wolff shipyard, but by the 1990s much of the area was disused and derelict.

Urban regeneration means taking action to try to bring new life to part of a city – improving the buildings, bringing in new employment and providing social facilities. An **urban planning scheme** is a large-scale project to improve an area in this way.

Evaluate the scheme

Benefits:
● The Titanic Quarter has 13,000 square metres of workspaces, and the fastest optical fibre link to the USA anywhere in Europe, and has attracted significant investment from large companies, bringing jobs to Belfast.
● There are 15,000–18,000 people currently living, working and studying in Titanic Quarter.
● The mix of housing, work and leisure in one area reduces problems of congestion and pollution as people do not have to commute.
● All the new buildings use energy conservation, solar heating and rainwater harvesting to reduce their environmental impact.

Costs:
● £385 million invested.
● Gentrification of an area which was previously a workplace for people on lower incomes.

Revision activity

Copy out the table and place each of the following statements in the appropriate column. Learn the four lists.

Improvements in the Titanic Quarter			
Housing	**Employment opportunities**	**Transport**	**Environment**

(a) 7500 apartments and townhouses to be built, such as the ARC, with 474 apartments.

(b) Premier Inn opened in 2010, Titanic Hotel in 2017 and a third hotel was approved in 2020.

(c) Derelict industrial site will be decontaminated (cleaned up).

(d) International firms Citigroup and Microsoft have opened offices here.

(e) Developers claim that 20,000 jobs will be created over fifteen years. In 2020 there were about 6000 new jobs.

(f) Public Records Office and Belfast Metropolitan College have relocated to Titanic Quarter.

(g) Former slipways where *Titanic* and *Olympic* were built laid out as parks and open space for residents and tourists, hosting large events such as Proms in the Park, the Titanic Maritime Festival, and drive-in events during the coronavirus pandemic.

(h) 1.5 km of attractive water frontage hosted the Tall Ships in 2009, 2015 and 2017.

(i) Titanic Studios is now one of Europe's largest film studios, hosting *Game of Thrones* and many others.

(j) Dedicated bus services, £5 million cycle path and pedestrian walkway upgrade at Lagan Weir opened in 2015, linking Titanic Quarter with the city centre.

(k) Titanic Belfast was named the leading visitor attraction in Europe in 2016; in 2019 it attracted 828,000 visitors, with over 6 million visiting since it opened in 2012.

Now test yourself

1 Name two companies that have located in Titanic Quarter.
2 How many houses and apartments are planned?
3 Give two ways the new buildings aim to protect the environment.
4 How much investment has been put into this planning scheme?
5 How many jobs are expected in the area?

Exam tip

This is a case study so you must know the details such as numbers and names. Make sure you know at least one detail for each of the four headings used in the table on page 82.

Exam tip

If you are asked to evaluate the scheme, you need to give the good and bad points, and then give an overall opinion on whether it has worked well or not. For example, after giving both sides, you could say 'clearly there are some problems associated with the regeneration, but overall it is bringing jobs and new life into an area which was previously completely derelict, so the positives outweigh the negatives'.

Revision activity

Make a postcard-sized summary of your case study.

Case study: Shanty town areas – Kolkata, India

People build their own houses because they cannot afford to rent or buy. These areas are called **shanty towns**. They are often unplanned, and lack basic services like clean water. In Kolkata the registered shanty towns are called bustees.

	In general	In Kolkata	Why?
Growth	Slow growth over long period of time, more rapid growth recently	Shanty towns have existed for 150 years. Now growing rapidly in number, they increased by 32% from 1981 to 1991, reaching 5500 by 2001. It is estimated that 4.5 million people now live in shanty towns, approximately one-third of the city's population	Historically, servants for the British rulers lived in shanty town areas. Many people have recently migrated to the city due to: Pull factors – job prospects in city Push factors – mechanisation in farming means people lose jobs in countryside People cannot afford to buy houses, or pay expensive rent, so they move to a shanty town, where they may rent or build
Location	Cheap land, edge of city, next to main roads, steep slopes, marshy land	City centre (older slums) such as Colutala Near factories and main road junctions, and along canals, such as Jagarani Vacant land, especially east of the city	Old buildings become derelict Build houses near factories for jobs, or roads for buses Cheap unwanted land, less likely to be bulldozed
Characteristics	Poorly constructed, often using scrap wood or corrugated iron, crowded, few facilities, no street plan, no sewage facilities	Defined as 'unfit for human habitation'. In registered bustees, people have the right to live there and slightly better conditions. Crowded: 10–12 people may share a room; 100 people share a tap, 25–30 people share a toilet. Average earnings £7–£24 per month – 75% of inhabitants are below the poverty line, 50% of children may be malnourished	Low wages mean people cannot afford 'normal' houses. Built by occupiers, quickly, to meet urgent needs. No planning permission. Electricity and water may only be provided years after building. 85% of shanty town residents have no sewage disposal; 22% have no steady income. The government cannot provide enough facilities for the number of people

Write out the following list of 'b' words, with at least one piece of evidence for each word. The first one has been done for you. Learn the 'b' words as a chant, and try to remember the evidence that goes with each one. Bustees are:
- big (4.5 million people in Kolkata, growing at 32% in ten years)
- badly built
- basic
- beside roads
- bulldozed
- below poverty line.

Now test yourself

TESTED

1 Other than push and pull factors, what explains the rapid growth of shanty towns in LEDC cities?

2 If farm workers move to Kolkata because more machinery is being used on farms instead of workers, is this a push factor or a pull factor?

3 If people move to Kolkata because of better job prospects, is this a push factor or a pull factor?

4 What term is used for shanty towns in Kolkata?

5 Approximately how many people live in Kolkata's shanty towns?

6 List all the words you can, beginning with the letter b, to describe Kolkata's shanty towns.

Exam practice

1 Describe what is meant by an urban planning scheme. [2]

2 For your case study of urban regeneration, describe and evaluate how successfully it has improved transport. [7]

3 Study the photograph below and answer the following questions.

(a) Explain why large numbers of people live in shanty towns. [3]

(b) Describe and explain the characteristics of shanty towns for one example of an LEDC city that you have studied. [6]

Exam tip

In a question such as 3b, check carefully which aspects of your case study they want – it could be location, **or** rapid growth, **or** characteristics of shanty towns, or they could ask you about two or even all three!

Revision activity

Make a postcard-sized summary of your case study. Check what you need to know. Make sure you include at least two pieces of detailed information, like place names or numbers, for everything.

Theme C Contrasts in World Development

The development gap

You need to be able to:

- identify and describe differences in development between MEDCs and LEDCs using social and economic indicators, with reference to places
- evaluate the use of social and economic indicators, and assess the advantage of using the Human Development Index (HDI)
- understand how the following factors can hinder development in LEDCs, with reference to places:
  - historical factors
  - environmental factors
  - dependence on primary activities
  - debt.

Differences in development between MEDCs and LEDCs

 REVISED

Development means the level of economic growth and wealth of a country. It affects people's income and their standard of living. For example, a more developed country is likely to have good healthcare and education systems, whereas a less developed country may not be able to provide these.

The development gap is the difference in development between the richer and poorer countries:

- MEDCs are More Economically Developed Countries, or richer countries such as the UK or the USA.
- LEDCs are Less Economically Developed Countries, or poorer countries such as India or Kenya.

Most MEDCs are in North America and Europe. Most LEDCs are in South America, Africa and Asia. Often a dividing line is drawn on a world map, showing richer countries mostly in the north (but including Australia and New Zealand!) and poorer countries mostly in the south.

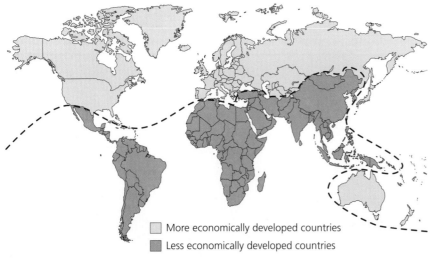

☐ More economically developed countries
■ Less economically developed countries

Figure 1 The north–south divide.

85

Development is difficult to measure. We use lots of different pieces of information to indicate how developed a country is. These are called development indicators. Some are to do with money – these are economic indicators, like how much money is earned. Others are to do with the way people live and their quality of life – these are social indicators, such as the percentage of people who can read and write.

Exam practice answers and quick quizzes at www.hoddereducation.co.uk/myrevisionnotesdownloads

How good are these indicators?

Economic and social indicators both have advantages and disadvantages.

Economic indicators, such as GNI per person:
- ✔ tell us how much money is available
- ✗ don't tell us what it has been spent on
- ✗ don't tell us whether everyone has the same amount of wealth
- ✗ don't tell us what people's lives are like
- ✗ might ignore food grown for a family, rather than to sell.

For example, a country with high GNI per person could have fantastic schools and hospitals, and everyone could be on good incomes. Alternatively, the money might all be spent on weapons and lots of people could be living in poverty, while a very few are rich.

Social indicators, such as average literacy figures:
- ✔ give us an idea of how money is being spent in a country
- ✔ show what people's lives are like
- ✗ don't show us the variety within the country. For example, they don't show whether there is inequality between boys' and girls' education
- ✗ don't show how much money is available for development.

If we only use one set of information, we only see part of the picture.

The Human Development Index (HDI) was created as a way to combine several pieces of information.
- ● Social: life expectancy, which tells us something about health.
- ● Adult literacy and school enrolment: which tells us something about education.
- ● Economic: gross national income per person as a measure of wealth.

Using HDI has several advantages:
- ✚ It gives us a picture of both wealth and quality of life, because it includes GDP and health and education measurements.
- ✚ Using education allows us to look at the potential for development in the future, as people with better education are likely to earn more money.
- ✚ HDI reveals how rich countries spend their money. For example, some countries which earn a lot of money from oil would have a high GDP figure. However, if the money is kept by a few very wealthy people, instead of being spent on healthcare and education, this will show up in poor literacy and life expectancy figures, which will give a low HDI figure overall.

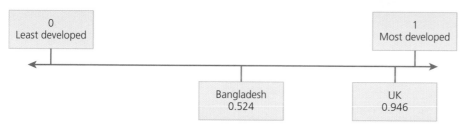

Figure 1 An HDI figure is always between 0 and 1.

Factors that hinder development in LEDCs

Something that 'hinders' development is something that slows it down or makes it more difficult. There are lots of different things which have hindered development in LEDCs. These can be divided into four categories:

- historical factors, things that have happened in the past
- environmental factors, things to do with the natural environment in the LEDC
- dependence on primary activities, meaning the country earns most of its money from primary activities, which are anything getting resources from nature, such as growing crops to sell, or mining
- debt, when a country borrows large amounts of money and has to pay it back with interest.

Revision activity

For each factor listed below, indicate whether it is (a) historical, (b) environmental, (c) dependent on primary activities, or (d) due to debt.

- Many LEDCs suffer hurricanes, earthquakes and floods.
- Some LEDCs have large debts, so spend their money on repayments instead of hospitals and schools.
- Ecuador owes $10 billion.
- In most LEDCs, large numbers of people work in primary activities such as farming or mining. This does not earn much money for the country.
- The 2004 Boxing Day tsunami destroyed large areas in India.
- European countries such as the UK and Spain in the past took over large areas of the world as colonies. They imported raw materials from the colonies, and the colonies did not get much money for this.
- Some diseases are common in hot wet climates, which mostly occur in areas with LEDCs.
- In the past the UK imported its cotton from India.
- Zambia gets 98% of its income from exporting copper.

Exam tip

You need to be able to illustrate your answer by talking about specific places, so learn the examples.

Exam practice

1 Classify the following indicators of development as social or economic by ticking the correct column. [3]

Social	Indicator	Economic
	Life expectancy at birth (years)	
	GNI per person (US$)	
	Literacy rate (%)	

2 Evaluate social indicators as a way of measuring development. [3]

3 The Human Development Indicator (HDI) was developed to try to measure development more effectively. Why might it be better than other indicators? [3]

4 Choose one of the following, and explain how it has hindered development in LEDCs: Historical factors, Debt, Environmental factors, Dependence on primary activities. You should refer to at least one place in your answer. [4]

Now test yourself

1 List the four types of factor that have hindered development in LEDCs.

2 Give one example for each type of factor to show how it has hindered development in one place.

TESTED

Sustainable solutions to the problem of unequal development

You need to be able to:

- describe how any three of the Sustainable Development Goals attempt to reduce the development gap
- define appropriate technology
- describe and evaluate the success of one appropriate technology product
- understand fair trade and the advantages it brings to LEDCs, with reference to places.

The Sustainable Development Goals

REVISED

The Sustainable Development Goals were adopted by the United Nations (UN) in 2015. They summarise what the UN would like to achieve by 2030.

The Sustainable Development Goals (SDGs) focus on improving quality of life, education and health as well as reducing poverty, inequality and negative environmental and climate change. There are seventeen goals, each with several targets. The goals are listed below.

1	No poverty	10	Reduced inequalities
2	Zero hunger	11	Sustainable cities and communities
3	Good health and well-being	12	Responsible consumption and production
4	Quality education	13	Climate action
5	Gender equality	14	Life below water
6	Clean water and sanitation	15	Life on land
7	Affordable and clean energy	16	Peace, justice and strong institutions
8	Decent work and economic growth	17	Partnerships for the goals
9	Industry, innovation and infrastructure		

Exam tip

Make sure you are familiar with this list – but you only need to know three in detail.

Now test yourself

TESTED

Categorise the Sustainable Development Goals under the headings below:
Environmental protection   Economic change   Social and cultural change

How do three goals attempt to reduce the development gap?

No poverty

One in five people are still in extreme poverty, living on less than $1.25 a day. This goal aims to eradicate poverty, so that people have higher incomes, and governments have more to spend on healthcare and education. Poor countries would become richer, reducing the development gap between the poor and rich countries.

Decent work and economic growth

The aim is to have full employment and decent work for all adults, including young people and those with disabilities. The poorest countries should have a growth rate of 7% a year in their gross domestic product (GDP). If this is achieved, these poor countries will rapidly become richer, reducing the development gap between the poor and rich countries.

Life on land

The aim is to manage forests sustainably, restore damaged forests, combat desertification and protect other ecosystems. This would mean countries can benefit from the diversity of the natural environment. This aims to prevent countries from ruining their environment in the rush to develop, and therefore should bring LEDCs in line with the requirements on MEDCs to conserve their environment for the future, reducing the gap between the poor and rich countries.

> **Now test yourself**
>
> 1 List three Sustainable Development Goals.
>
> 2 For each one, explain how it attempts to reduce the development gap.
>
> TESTED

Appropriate technology

Technology is the method or tool which is developed to carry out a task. Appropriate technology is technology which is appropriate to the situation. For example, it would be no good giving tractors to farmers in poor countries if they cannot get the fuel to run them – it wouldn't be appropriate.

Appropriate technology usually involves small-scale technology which can be controlled and maintained by local people, using local resources, at low cost.

> **Now test yourself** TESTED
>
> Copy the table. Place a tick in the column by the options which are most likely to be appropriate for LEDCs.

	Option (a)	✔	Option (b)	✔
1 Fuel	Uses solar power		Uses coal and oil	
2 Materials	Uses wood from local forests		Uses steel from USA	
3 Maintenance	Local people know how to maintain equipment		Needs lots of experts from Europe	
4 Cost	Costs less than a day's wages		Costs three months' wages	
5 Environmental impact	Causes lots of pollution		Causes very little damage to environment	
6 Jobs	Machinery is made in France		Machinery can be made in the village	

Evaluate one appropriate technology product

The Hippo water roller is a 90-litre barrel for carrying water which can be put on its side and rolled along using a handle.

Figure 2 The Hippo water roller.

Its benefits include:
- ✔ requires no spare parts
- ✔ requires no power
- ✔ durable – lasts up to ten years
- ✔ barrel can be recycled
- ✔ speeds up fetching water, giving women and children more time for employment or education
- ✔ reduces long-term neck and spinal damage from carrying heavy water containers
- ✔ used in more than twenty African countries, providing water for more than 300,000 people.

However, there are some disadvantages:
- ✗ it costs $90 per barrel, so is dependent on donations
- ✗ if the barrel is not full it can be difficult to control
- ✗ the barrel is difficult to fill from a shallow source of water
- ✗ mud in the water doesn't settle to the bottom the way it would in a traditional water container, as it is permanently moving around.

Now test yourself

TESTED

1. How much water can the Hippo roller carry?
2. How long does it last?
3. Give three features that make it appropriate for LEDCs.
4. How many countries is it used in?

Fair trade

Fair trade means people who make or grow something are paid a fair price for their work. This price is guaranteed, so the producer will not lose out if world prices fall.

If you buy a bar of chocolate which is *not* fair trade, the profits are shared as shown in Figure 3. Fair trade means the producers get more of the money.

The advantages of fair trade for LEDCs:
- guarantees a minimum wage for farmers
- farmers can provide for their families
- farmers have access to cheap loans
- farmers control the business
- profits are used by groups of farmers to help provide healthcare, education and transport for their communities
- encourages sustainable farming practices.

Places that have benefited from fair trade are St Lucia and Tanzania:
- The island of St Lucia in the West Indies has thirteen groups of fair trade banana farmers. Since 2000 they have used part of their income from selling fair trade bananas to build a community centre, provide science equipment in two schools, and buy a new truck to deliver fertilisers and packaging materials to members.
- 60,000 coffee growers in Tanzania have a cooperative union which helps them to sell fair trade coffee. This helps to improve housing and provide education.

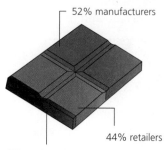

52% manufacturers

44% retailers

4% producers

Figure 3 Profit distribution of a chocolate bar without fair trade.

> **Exam tip**
>
> You need to be able to write about places, so make sure you can talk about two places which have benefited from fair trade.

Exam practice

1 Name three of the Sustainable Development Goals introduced in 2015. [3]

2 State the meaning of the term 'fair trade'. [2]

3 Study the graph which shows changes in the market price of coffee.

Describe the changes in the price of coffee on the world market shown in the graph. [3]

4 Study the table which shows the price paid for coffee on the world market and the price paid by Café Direct (a fair trade company) in 2005. Explain how Café Direct helps coffee growers. [3]

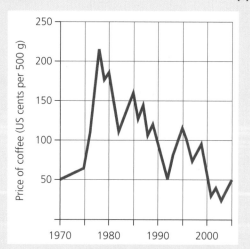

	World market price (US cents per 500 g)	Café Direct pays (US cents per 500 g)
Arabica coffee	120	132

5 **(a)** Describe one appropriate technology product. [3]
   (b) Evaluate the product. [6]

> **Revision activity**
>
> Draw a coffee jar or bar of chocolate and label it with:
> - bad things about normal trade in black
> - good things about fair trade in green
> - two places which have benefited from fair trade in blue.
>
> Each time you eat chocolate, drink coffee or eat a banana, repeat the advantages of fair trade!

> **Exam tip**
>
> For full marks in question 3, you need to refer to the price rising until 1978 and then falling unevenly, as well as quoting figures from the graph.

Globalisation

What is globalisation?

REVISED

The diagram below describes what is meant by globalisation.

World trade – brands can be sold everywhere – you can buy Coca-Cola anywhere

Countries sell to each other and are therefore **interdependent** – they depend on each other

What is globalisation?

The way people, goods, money and ideas move round the world faster and more cheaply than ever before

Industries have become global – individual companies operate in lots of countries as multinational corporations (MNCs). These are very powerful

Economic decisions or **events in one country affect other countries** quickly

Figure 4 Globalisation is …

Now test yourself

TESTED

1 Give a definition of globalisation.
2 Give three different aspects of globalisation.

How can globalisation both help and hinder development?

REVISED

When a multinational corporation (MNC) sets up a factory in an LEDC, it provides jobs. This gives people an income to improve their standard of living. They buy more in shops, meaning other people who own those shops, or make things for those shops, make more money. They pay taxes, giving the government more money to spend on development, such as healthcare or education.

However, many of these jobs are poorly paid, with low job security, as the MNC could move their factory somewhere else if it becomes cheaper. Most of the profits go to the MNC headquarters, which are usually located in MEDCs.

93

Case study: How has globalisation affected India, a BRICS country?

BRICS refers to **B**razil, **R**ussia, **I**ndia, **C**hina and **S**outh Africa, which were all experiencing similar rapid industrial growth in 2001.

Since the 1990s, entrepreneurs have been encouraged to set up businesses in India. Lots of MNCs have invested in India. There are lots of people who are seeking work and will accept low wages. Large numbers of people speak English, and lots of people who previously emigrated from India to get work in other countries are now returning with new skills.

Revision activity

Evaluate the impacts of globalisation in India. Decide whether the following are evidence of how globalisation is helping development, or hindering it. Make two separate lists, and learn them:

1 Half of children under five years of age in India are malnourished.

2 Life expectancy has gone up from 59 years in 1990 to 68 in 2015.

3 More people in India now have cars, TVs, washing machines and other consumer goods.

4 More imported goods mean there are fewer jobs in factories for those with little education.

5 MNCs have created many new jobs in call centres and hi-tech industries.

6 Western-style clothes and behaviour are considered shocking by some.

7 Adult literacy rates have increased from 50% in 1990 to 74% in 2011.

8 Increase in the number of enormous shopping centres.

9 Unrest in rural areas has led to guerrilla fighting.

10 Less than one-third of homes in India have a toilet, and less than half the villages are connected to the electricity network.

If you have a different case study, make sure you have two separate lists for it – one for how globalisation has helped development and one for how globalisation has hindered development – and learn them!

Now test yourself

TESTED ◯

1 Give two pieces of evidence that globalisation has helped development in India.

2 Give two pieces of evidence that globalisation has hindered development in India.

Exam practice

1 State the meaning of the term 'globalisation'. [2]

2 Explain what is meant by a BRICS country. [3]

3 For one case study, explain how globalisation has affected development. [7]

Revision activity

Make a postcard-sized summary of your case study. Check what you need to know. Make sure you include at least two pieces of detailed information, like place names or numbers, for everything.

Exam tip

In a case study question, you should aim to give at least two specific details – such as place names or numbers. These need to be relevant to the question, so for some case studies you will need to learn several. For example, here you could get a question about how globalisation helps development in your case study, so you would need two details about the positive side of globalisation. Alternatively, they could ask how it hinders development, so you would need two details about the negative side.

Theme D Managing our Environment

Human impact on the environment

You need to be able to:

● describe the greenhouse effect, define a carbon footprint and explain how both of these can contribute to climate change
● evaluate the effects of climate change on the environment, people and the economy, with reference to places.

The greenhouse effect

The greenhouse effect is a natural process, where heat from the Sun is trapped in our atmosphere by greenhouse gases such as carbon dioxide (CO_2) and methane (CH_4). It keeps the Earth warm enough for us to live on it. Without the greenhouse effect, the planet would be freezing and nothing could survive.

Exam tip

Think of it like being in a greenhouse, or a classroom with lots of windows and the Sun shining in. The Sun heats the air, and then the windows keep the heat in, so the classroom or greenhouse gets warmer.

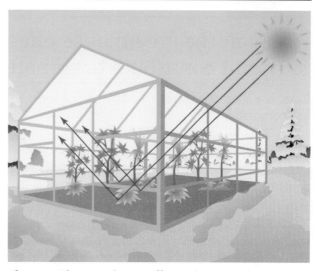

Figure 1 The greenhouse effect … in a greenhouse!

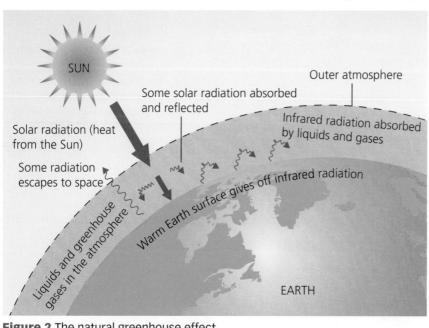

Outer atmosphere

Some solar radiation absorbed and reflected

Infrared radiation absorbed by liquids and gases

SUN

Solar radiation (heat from the Sun)

Some radiation escapes to space

Liquids and greenhouse gases in the atmosphere

Warm Earth surface gives off infrared radiation

EARTH

Figure 2 The natural greenhouse effect.

95

What is a carbon footprint?

A carbon footprint is a measurement of the amount of greenhouse gases produced by a person, organisation or activity in a given time such as a year. All the gases are converted into the equivalent weight of carbon dioxide to make it simpler to compare how harmful different activities are.

For example, travelling by car means that petrol is burnt. This comes from oil, which is a fossil fuel made from dead plants and animals. These living things absorbed carbon while they were alive, and it is trapped in the oil. When we burn the oil, it releases the carbon dioxide into the atmosphere. Many products are transported by road, so when you buy them, this increases your carbon footprint.

Lots of our electricity is produced by burning oil or gas (another fossil fuel) to boil water and then using the steam to turn a turbine, which generates the electricity. This means whenever you use electricity you might be increasing your carbon footprint.

Exam tip

Be ready to explain an example in full: going on holiday to Spain in a plane requires aviation fuel to be burnt. This comes from oil, which is a fossil fuel made from dead sea creatures. When aviation fuel is burnt, it produces carbon dioxide, so this increases a passenger's carbon footprint.

Now test yourself

How could the following increase your carbon footprint?

1 Travelling to school by car.
2 Going on holiday to Spain.
3 Buying grapes grown in Italy.
4 Turning the heating up in your house.
5 Using a hairdryer to dry your hair.

TESTED

How can the greenhouse effect and carbon footprint contribute to climate change?

Climate means the average weather over 30 years. This includes rainfall and temperature.

Climate change is the idea that the climate of different parts of the world is changing. It seems that the world as a whole is getting warmer, but it could change other things too – for some places it will be drier, and for others it could be wetter.

If the amount of greenhouse gases increases, it is like putting a thicker blanket around the Earth – it will keep more heat in. So, if people or organisations have big carbon footprints, this will contribute to climate change by increasing the greenhouse effect.

Exam tip

You might need to go one step further. The question could ask how going on holiday to Spain could contribute to global warming.

Use the answer above, and add:

Carbon dioxide is a greenhouse gas. When the Sun heats the Earth, warmth is trapped by greenhouse gases. When human activity adds greenhouse gases to the atmosphere, more warmth is trapped. This results in the globe warming up, creating changes in the climate.

Now try the same thing for the other examples.

Now test yourself

TESTED

Caroline Careless has a carbon footprint of 12.56 tonnes but Gregory Green's footprint is only 2.00 tonnes. The table lists some of the choices they each have made about how they live. Study the list and decide who made each of the choices listed – tick either Caroline or Gregory.

Lifestyle choice	Caroline	Gregory
(a) Lives in a house which is poorly insulated		
(b) Only buys low-energy light bulbs		
(c) Has solar panels on the roof to heat water		
(d) Always leaves the TV and computer on stand-by		
(e) Hangs the washing to dry on a clothes line outside		
(f) Washes clothes on a cool wash at 30°C		
(g) Drives one mile to the shop to buy milk		
(h) Always drives instead of travelling by bus or train where possible		
(i) Walks or cycles to local shops		
(j) Goes on long-haul flights for two holidays every year		

Evaluate the effects of climate change

Revision activity

The following statements are descriptions of effects that climate change could have. For each, decide if it is positive or negative, and whether it is an effect on the environment, people or the economy. Some may be more than one!

Effect	Positive or negative?	Is this an effect on the environment, people or economy?
(a) The **Arctic Ocean** could lose all its sea ice in summer within twenty years. The dark seawater will absorb more of the Sun's rays, making other areas melt more quickly		
(b) Food production could be reduced, and providing water will be expensive – possibly making the world $3.88 trillion poorer		
(c) Rising sea levels could cause floods in low-lying areas such as near **Strangford Lough**, destroying wildlife habitats and farmland		
(d) 20–30% of species could become extinct		
(e) Warmer air contains more moisture, bringing more rain – in the **UK**, 2012 was the wettest on record. The cost of dealing with flood damage could increase from £1 billion to £12 billion per year.		
(f) Warmer oceans pick up more CO_2 and become acidic. This harms coral reefs such as the **Great Barrier Reef in Australia**, as it 'bleaches' them, causing 1500 km of coral death already		
(g) If nothing is done about climate change, **India** is forecast to have a 92% reduction in gross domestic product by 2100		
(h) Warm summers could lead to deaths in droughts, as food supplies are affected, and in heatwaves		

Now test yourself

TESTED

1 How much is India's GDP expected to fall by?

2 How much of Australia's Great Barrier Reef has been bleached?

3 Name a place that would be affected by sea level rise.

4 If sea ice in the Arctic Ocean melts, what will happen to other snow and ice in the area and why?

Exam practice

1 Explain what is meant by the greenhouse effect, and how it can contribute to climate change. [4]

2 Study the poster on the right. Choose one of the ideas in the poster. Explain how it would help reduce your carbon footprint. [3]

3 Evaluate the effects of climate change on the economy. Refer to places in your answer. [5]

Exam tip

You need to write about places for this topic: so learn one fact about each of the places in bold in the table.

5 ways to reduce your CARBON footprint!

1 **Walk or cycle**
2 **Turn off lights when not needed**
3 **Don't leave the TV on standby**
4 **Don't buy so much bottled water**
5 **Eat locally grown food**

Strategies to manage our resources

Waste management

REVISED

Waste management is how litter and other waste is dealt with. For a long time, waste in Northern Ireland has mostly gone to landfill sites, where it is simply buried underground in old quarries or natural dips in the ground.

This is a problem because:
● We are running out of landfill sites – lots are full already, and people don't like new ones near where they live.
● People worry that chemicals from the waste will poison groundwater and spread disease.
● There are strict government targets. By 2035, the UK aims to have only 10% of waste going to landfill.

The waste hierarchy

There are lots of ways of dealing with waste which are more sustainable than landfill. The waste hierarchy lists these from most sustainable at the top, to least sustainable at the bottom. The most sustainable, reducing, is widest in the diagram, as the government wants that to be what happens most, while disposal in landfill is narrowest as it is least sustainable, so it should be what happens least.

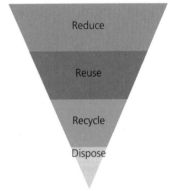

Figure 3 The waste hierarchy.

Reduce, reuse, recycle

There have been campaigns to encourage people to use the most sustainable of these options: reduce, reuse, recycle.
● 'Reduce' means produce less waste, for example by buying less, or choosing products that have less packaging.
● 'Reuse' means you can use an item again as it is, such as passing things on to a friend, or donating them to a charity shop.
● 'Recycle' means an item can be turned into something else, such as plastic bottles being turned into a park bench. This uses energy, so it is not quite as sustainable as the other options.

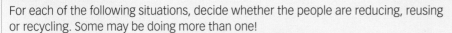
For each of the following situations, decide whether the people are reducing, reusing or recycling. Some may be doing more than one!

(a) Gill's dad always keeps paper that has been printed on one side to use as scrap paper.

(b) Jenny uses a 'bag for life' when she goes shopping, instead of getting plastic bags each time.

(c) Barry the Builder always takes broken metal from old buildings to the scrapyard where it will be melted and turned into new metal.

(d) Oliver always takes his lunch to school in a lunch box so he doesn't need clingfilm.

(e) Theo puts all the empty cans in the blue bin instead of the grey one.

(f) Mum bought low-energy light bulbs for the whole house. She said they should each last six years.

(g) Deborah's new pencil case was made from an old car tyre.

(h) Susan's baby wears washable nappies instead of disposable ones.

Revision activity

Draw the waste hierarchy and label it correctly. Next to each layer, add detail to explain what it means, and explain what makes it sustainable or unsustainable.

A renewable energy source

What is a renewable energy source?

Coal, oil and natural gas are non-renewable (or finite) energy resources. The more of them that are extracted from the Earth and used, the fewer energy reserves there are left for use by future generations. This is not **sustainable**. Burning coal, oil and gas also contributes to climate change and acid rain so that the environment is damaged for the future too. Using renewable energy sources, which will not run out, is therefore more sustainable.

Some renewable energy sources are:
● Solar energy – where the Sun's energy is used to create heat or electricity.
● Wind energy – where the wind turns a turbine, which generates electricity.
● Biofuels – where plants such as willow trees, maize or sugar crops are burnt or processed to create energy.

How is solar energy used?

Method	How it works
Concentrated solar power (CSP) – parabolic troughs	Curved mirrors focus the Sun's rays onto a tube of liquid, which is heated to very high temperatures, to create steam, which turns a turbine and generates electricity
Concentrated solar power (CSP) – power tower system	Flat mirrors are controlled by computers to follow the Sun all day, to focus its rays on a tower in the centre, heating a liquid, which creates steam, turns a turbine and generates electricity
Photovoltaic cells	Sunshine creates chemical changes, which generate electricity. This is like a solar-powered calculator, only much bigger

Benefits and disadvantages of solar power

For each of the following, decide whether it is a benefit or a disadvantage, and complete the table by ticking the appropriate column.

Solar power	Benefit?	Disadvantage?
Creates **no pollution**, noise or greenhouse gases		
Difficult and expensive to generate electricity for **night**-time use		
Less production in **winter**, cloudy weather or storms		
Cheap to run – no fuel needed		
Photovoltaic cells last about **40 years**		
Photovoltaic cells are only **20% efficient**, so large areas are needed		

Revision activity

Use the words in bold to create two lists which you can learn to jog your memory – one list of benefits and one of disadvantages.

Exam tip

If you are asked to evaluate a form of renewable energy, you need to give both good and bad points, then an overall opinion. For example, you might conclude 'although there is work needed to make solar power methods more efficient, they are still beneficial as they produce electricity without pollution'.

The 2015 Climate Change Agreement

REVISED

Description

- **What is it?** An agreement between 195 countries to reduce their greenhouse gas emissions.
- **Why?** To try to prevent global temperatures increasing by more than **1.5°C**, to avoid lots of problems such as sea level rise and droughts.
- How? The United Nations spent nine years negotiating with different countries. It is sometimes called the **Paris Agreement**, because it was finally agreed at an international conference held in Paris. Each country has a **target** for reducing their greenhouse gas emissions, which will become more ambitious every few years.

Revision activity

Make yourself a revision card with the words in bold, then learn them. Practise using those words to help you describe the Agreement.

Evaluation

Complete the table by ticking the correct column. Some things may be both positive and negative – make sure you can explain why!

2015 Climate Change Agreement	Positive	Negative
Each country has to announce its progress every five years		
No outside agency to make sure countries stick to the Agreement – they control themselves. The USA pulled out of the Agreement from 2017 to 2021		
Airlines can plant trees to absorb the greenhouse gases they produce, eliminating 2.5 billion tonnes of emissions		
LEDCs are given more time to reduce their emissions as they need to raise living standards first		
It will cost money for countries to reduce their emissions		

Exam practice

1 Study the graph below.
   (a) Which city has the least sustainable waste disposal? Give evidence from the graph. [3]

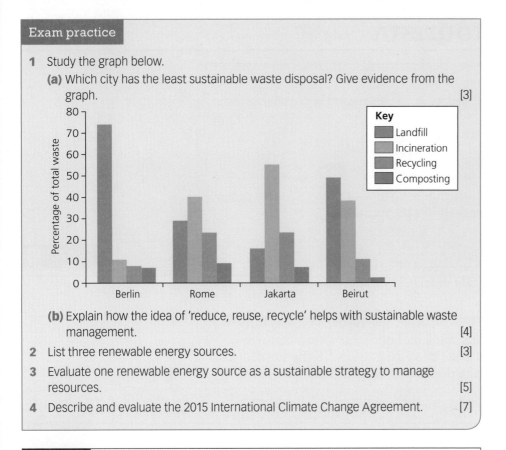

(b) Explain how the idea of 'reduce, reuse, recycle' helps with sustainable waste management. [4]

2 List three renewable energy sources. [3]

3 Evaluate one renewable energy source as a sustainable strategy to manage resources. [5]

4 Describe and evaluate the 2015 International Climate Change Agreement. [7]

Exam tip

In an exam question, if you are asked to evaluate, you need to give both the positives and negatives, and then an overall opinion at the end – for example, you might say 'overall, I think the Agreement is good, and will be effective at reducing carbon emissions, but there may be some problems with implementing it'.

Sustainable tourism

You need to be able to:

- evaluate the cultural, economic and environmental impacts of mass tourism, with reference to places
- describe and explain how to be a responsible tourist
- describe and explain ecotourism
- assess how ecotourism can protect the environment, using a case study.

Sustainable tourism is tourism which tries to protect the local environment and people, without causing problems for the future.

Mass tourism

Mass tourism means large-scale tourist developments for large numbers of tourists.

Tourism has increased since the 1960s. In the 1960s, a typical family would feel fortunate to spend a week on holiday at a seaside town in this country. Since then, there has been an increase in the number of holidays taken per year and the distance travelled. There are a few reasons for this:

● **Increased leisure time.** Workers now get more days' annual holiday so may take short-break holidays as well as the usual 'summer holiday'.	● **Increased disposable income.** The incomes of workers have grown faster than the cost of living. Most people now have a greater 'disposable income': the money they can choose to spend, once they have paid for the essentials. This means many people can afford to travel much further than in the past.
● **Cheaper travel.** Low-cost airlines such as easyJet and Ryanair have dramatically reduced the cost of flying. This makes weekend breaks in foreign cities more affordable, and has encouraged some people to buy second homes in Mediterranean countries, travelling cheaply and conveniently several times per year. Cheap flights also encourage travellers to plan longer journeys that would once have seemed too expensive.	● **Increased health and wealth of pensioners.** Older people are able to travel more than in previous generations as they are living longer and in better health. They are not restricted to popular school holiday times so they can take advantage of cheaper off-peak holiday prices. If they have saved for their retirement they can sometimes afford several holidays a year.

All this means many areas are now affected by mass tourism, especially Spain, Greece and Portugal.

Impacts of mass tourism

Impacts	Positive	Negative
Economic	Jobs created in hotels, restaurants, taxis, souvenir shops People employed in tourism spend their income in the local shops and services, boosting the local economy	Money earned from tourism is often lost to the local economy because, for example, hotel profits are sent to the owners overseas. If tourism uses migrant workers, they may send some of their wages home so local spending is reduced
Environmental	New airports and better roads aimed to benefit tourists actually improve the environment for everyone Local councils often take better care of historic buildings and scenic countryside when they know these attract tourists	Water is used in vast quantities for swimming pools and golf courses, and may leave a shortage for farmers and local people Wildlife and ecosystems can be endangered by the litter, sewage and other pollution created by tourists
Cultural	Both visitors and local people get the chance to learn from each other's culture. Tourism may bring in money which can help preserve traditional cultures	Local young people can become involved with drunkenness and crime associated with some tourist resorts Traditional costumes and dancing become detached from their genuine culture and devalued when performed to earn money from visitors Some tourists' behaviour, such as naked selfies, may offend local people

Now test yourself

TESTED

Read the following list of tourism impacts carefully. For each one, decide whether it is positive or negative and whether it is economic, environmental or cultural.

Impacts of tourism	Positive or negative	Economic, environmental or cultural
In **Cyprus**, when turtles hatch at night from eggs laid on beaches, they should creep towards the moonlit sea, but are endangered by bright hotel lights which lure them towards roads		
Local residents in **Zanzibar** use less than 50 litres of water per day, but tourists in hotels use 931 litres each per day on average		
Over 200 million people are employed in tourism worldwide, earning 11% of global earnings		
38% of visitors to **Uluru (Ayers Rock)** climb to the top even though the rock is sacred to the Aboriginal people of central Australia who request visitors not to make the climb		
An estimated 17% of **Kenya**'s tourist earnings 'leaks away' from the Kenyan economy		

Exam tip

You need to be able to write about places in the exam to illustrate your point. Add the examples from the list in the 'Now test yourself' section to the correct box in the table at the top of the page. Then make sure you learn the examples so that you can use them to illustrate the impacts of tourism.

How can you be a responsible tourist?

A responsible tourist is a tourist who respects the environment and people in the place they visit.

Read the story below, and pick out four ways in which Sonia and Trevor were responsible tourists. There are also four examples where they could have been much more responsible – explain how.

Sonia and Trevor were going on the holiday of a lifetime to Africa. Trevor did some research to make sure they could say 'please' and 'thank you' in the local language. When they arrived at their hotel they were really hungry, so they went straight out to McDonald's as they were glad to see a familiar logo. When they got back they met some ladies selling their homemade jewellery outside the hotel, and they bought some lovely elephant tusk necklaces. They also took selfies with the ladies, but they asked permission first.

The next day they went on safari, and picked some amazing flowers by the side of the road. They saw giraffes close up, and they were really careful to make sure none of the wrappers from their picnic blew away. When they got back to the hotel they were hot and tired, but after a long shower they went to the local restaurant and tried some local dishes – they were amazing.

Ecotourism

Ecotourism is where visitors enjoy nature at first hand while protecting the environment and local way of life. It is a bit like responsible tourism, but it is usually specifically in areas of natural environment, whereas responsible tourism could be in a city.

Figure 4 Ecotourism: a tour guide takes a photo of a tourist helping at a farmer's rice field in Bali.

How can ecotourism protect the environment?

Case study: Ecotourism in Nam Ha, Laos

Revision activity

Make a large copy of the table below and choose the appropriate statements from the following list to fill each box. Some statements may belong in more than one box. If your case study is different, fill out the boxes with facts about it, and learn them.

(a) Laos – an LEDC in South-East Asia (population 6.8 million).

(b) Ecotourism project organised by UNESCO.

(c) Wilderness area of mountains and deciduous forests.

(d) Guides and trekkers help to deter poachers so rare species are conserved.

(e) Nam Ha National Protected Area in the north of Laos.

(f) Major attractions include boating on Nam Ha River and trekking through forests.

(g) All trekking and boat trips must use Nam Ha Eco-guide Service.

(h) Wildlife here includes rare clouded leopards and tigers in danger of extinction, gibbons, Asian elephants and 288 bird species.

(i) Two proposed roads, which would have led to logging and illegal trade in wildlife, have not been allowed to go ahead.

(j) Local people become guides instead of hunters so wildlife is conserved.

(k) Visitor numbers have increased from 1000 to 4000 a year.

Facts about the location	
Description of the ecotourism project	
How it has protected the environment	
Problems?	

Exam practice

1. Mass tourism has increased significantly since the 1960s. Explain two reasons why this has happened. [4]

2. Evaluate the economic impact mass tourism has had, with reference to places. [5]

3. Study the information below.

> Uluru is one of Australia's most recognisable natural landmarks. It stands at 863 m high and has a total circumference of 9.4 km. It was named Ayers Rock by European explorers in 1872 but is now more commonly known by its original, Aboriginal name.
>
> The rock is sacred to the Anangu, the local Aboriginal people, and the Australian government handed back control of the area to them in 1985.
>
> The Anangu do not climb Uluru because of its great spiritual significance and they request that visitors should also not. Despite the Anangu's wishes, and the many dangers, about a third of visitors still climbed Uluru and it had to be banned in 2017.

Does this account show responsible tourism? Use evidence to explain your answer. [4]

4. With reference to a case study, assess how ecotourism can protect the environment. [7]

Revision activity

Make a postcard-sized summary of your case study. Check what you need to know. Make sure you include at least two pieces of detailed information, like place names or numbers, for everything.

Exam tip

In question 3, you will need to define responsible tourism, then pick out evidence from the resource. Try not to copy sentences – explain in your own words.

Exam tip

In question 4, explain what is meant by ecotourism and give examples of the ecotourism development in Laos, with some case study detail. Then give two ways it has helped protect the environment, with case study detail.

Fieldwork

You are allowed to take information into this exam with you. You should have a statement, which includes your title, aim and two hypotheses, and the details of where you did your fieldwork.

You also need a table of your data. This must include all the data relevant to your aim. It can include secondary data, and some data must be numbers so that you can draw a graph. It needs a clear title, and the columns must be clearly labelled, including the units of measurement.

Both of these must be created using information and communications technology (ICT). Your teacher will probably collect these in advance, and give them out on the day of the exam.

Exam tip

Keep a copy in your own file, so that you can read it over and practise your answers.

Exam tip

Check your spelling! One hypothesis but two hypotheses – just like one axis but two axes.

The geographical enquiry process

You need to be able to:

- explain the geographical enquiry process.

This means explaining how we carry out an enquiry from start to finish.

Exam tip

On the front page of the exam paper will be a statement that the quality of written communication will be assessed in a particular question. It might be worth putting a star against that question at the start of the exam, so that when you reach it you remember.

Don't let this worry you too much – it just means the examiner will check that specific question for spelling and grammar, and will want to see some specialist terms used. Concentrate on getting the information right, but take a minute afterwards just to check it over for spellings.

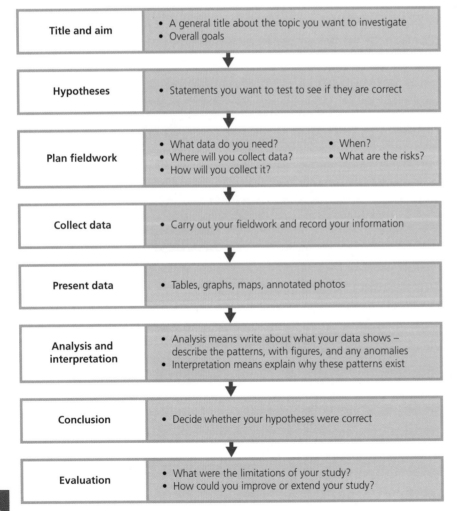

Title and aim	• A general title about the topic you want to investigate • Overall goals
Hypotheses	• Statements you want to test to see if they are correct
Plan fieldwork	• What data do you need? • When? • Where will you collect data? • What are the risks? • How will you collect it?
Collect data	• Carry out your fieldwork and record your information
Present data	• Tables, graphs, maps, annotated photos
Analysis and interpretation	• Analysis means write about what your data shows – describe the patterns, with figures, and any anomalies • Interpretation means explain why these patterns exist
Conclusion	• Decide whether your hypotheses were correct
Evaluation	• What were the limitations of your study? • How could you improve or extend your study?

Now test yourself

1 For each description below, decide which part of the enquiry process it fits into, then put the statements in the correct sequence.

(a) Brendan drew a graph to show all his data about river depth.

(b) Jessica had enjoyed what she had learned about managing the environment the most, so she wanted to investigate waste disposal and recycling in Belfast.

(c) Ben found a lot of patterns in his data, for example, as they travelled further from the source of the river, the stones on the river bed got smaller. He remembered to use some figures to demonstrate what he found, and he could explain it because he had learned about processes of erosion like attrition and abrasion.

(d) When Hannah looked at her data she decided that her first hypothesis was correct, but the second was not. She used her evidence to prove this.

(e) Niamh decided that she would carry out her fieldwork on Tuesday, in the main street, and she wrote a questionnaire to help her find out how far people had travelled to get there.

(f) Zeke realised that his fieldwork was limited because he had only visited one river, on one day, after a heavy rainstorm. It would be better to repeat the study several times in different weather conditions to get a better idea.

2 One stage in the enquiry process is missing above. What is it? Make up a statement for the missing stage.

Now test yourself

1 Give the missing stages in the sequence:
   - Title and aim.
   - Hypotheses.

   - _____

   - Collect data.

   - _____

   - _____

   - Draw a conclusion.
   - Evaluate the process and the conclusion.

2 What is a hypothesis?

3 Give two things you need to include in your analysis.

Exam practice

1 When planning fieldwork, it is important to decide where to collect your data. Give two other things that should be included when making fieldwork plans. [2]

2 After collecting your data, describe two other stages in the enquiry process. [4]

Planning (including aims and hypotheses)

Exam tip

For every section, you will mostly be asked about **your** fieldwork – knowing something in theory is not enough, you have to be able to explain what **you** did.

You need to be able to:

- explain how you planned your enquiry by:
  - identifying questions or issues for investigation;
  - developing **one** aim; and
  - developing **a minimum of two** appropriate hypotheses
- recognise the potential risks involved in **your** fieldwork and how to reduce these risks
- explain the difference between primary and secondary sources.

Questions, aim and hypotheses

REVISED ⬤

Questions for investigation are very broad. You narrow these down and decide on one aim – a summary of what you are trying to find out. Then you make at least two hypotheses, or statements, which you think are true, but you are going to test. Your hypotheses need to help with your aim, and you need to be able to explain how they do so. An example is given below.

- **Topic** – river environments.
- **Question** – how do rivers change downstream?
- **Aim** – to test the Bradshaw model.
- **Hypothesis 1** – river cross-sectional area will increase downstream – this helps with the aim, because the Bradshaw model says width and depth will increase downstream, due to increased water volume and erosion processes such as abrasion, which means cross-sectional area will also increase. If this hypothesis is correct it helps to confirm the Bradshaw model.
- **Hypothesis 2** – load particle size will decrease downstream – this helps with the aim, as the Bradshaw model predicts that the particle size will decrease, due to attrition, so if this hypothesis is correct it helps to confirm the Bradshaw model.

Exam tip

Make sure you can write about your own fieldwork like this. Remember, the statement you take in with you includes the question/issue, aim and hypotheses. Your answer must use the same hypotheses as your statement!

Now test yourself

TESTED ⬤

What is the difference between an aim and hypotheses?

Exam tip

Check ahead in the exam paper! You will probably be asked to use two of your hypotheses at different times. If you use the same hypothesis when the exam asks for a different one, you will lose marks. Check the questions involved so that you can plan the best hypothesis for each one. For example, if you want to draw a graph based on your first hypothesis, make sure you use your second hypothesis for the other question. Plan this at the start of the exam, so that you don't end up using the wrong hypothesis by mistake.

Risks and reducing them

REVISED ⬤

When you carry out fieldwork there are always some risks – the possibility that something could go wrong and cause someone harm. Teachers have to carry out full risk assessments, but the examiners want to know that you have thought about risk as well.

You should be able to write about three risks, explaining exactly what the danger was for **your** fieldwork, and what **you** did to reduce the risk.

For example, when Niamh was carrying out a survey about the distance people travelled to a town, there was a risk of getting run over, because of busy traffic, and of danger from talking to strangers. To reduce the risk, she chose a place with a wide footpath, to avoid blocking the path and making people more likely to walk in the road. She used pedestrian crossings so that she could cross the road safely. She worked with a partner, and only worked in the main street where there were plenty of people around, rather than in quiet side streets, as the presence of other people would help to deter any potential criminal activities.

> **Revision activity**
>
> Make a revision card, with three risks from **your** fieldwork on one side, and something **you** did to reduce each of those risks on the other side.

Primary and secondary sources

REVISED

A primary source is information you have collected yourself by doing fieldwork. This might be:

- quantitative, where you collect numbers, such as river width or numbers of cars
- qualitative, where you describe something, or give opinions.

A secondary source is information other people have collected. This could be from places such as the internet, or a textbook, newspaper or map.

> **Exam tip**
>
> The examiners will accept census data as either primary or secondary data.

You need to be able to explain what primary and what secondary data **you** used. For the secondary data, you should be able to explain how it helped with your investigation.

For example, Amid used a map showing vegetation on a sand dune to help decide where to carry out his fieldwork, as he wanted to check the natural vegetation, so he needed to avoid the area where people had planted a coniferous woodland.

Now test yourself TESTED

For each of the following, decide whether it is primary or secondary. Some could be either!

(a) measuring the depth of a river

(b) carrying out a survey asking people questions

(c) reading a newspaper article about tourism and how it affects Portrush

(d) identifying land uses for buildings along a road

(e) using the Northern Ireland Neighbourhood Information Service (NINIS) to collect census data about deprivation in Ballymena

(f) giving a place a score for its environmental quality.

Revision activity

Make a revision card listing two secondary sources you used in one colour, and how they helped you in another.

Exam practice

1 State the aim of your investigation, and explain how one of your hypotheses helps you achieve your aim. [4]

2 For **your** fieldwork, give two risks you identified, and explain what measures you took to reduce the risk. [4]

3 Name one secondary data source **you** used in your investigation, and explain how it helped. [4]

4 Explain why **your** fieldwork location was appropriate for your study. [3]

Fieldwork techniques and methods

You need to be able to:

- select data collection methods and equipment that ensure accuracy and reliability
- record measurements and observations accurately using recording sheets
- use at least one secondary source during the fieldwork investigation.

Selecting methods and equipment

REVISED

You could be asked to describe the data collection methods and equipment that **you** used, and how they helped you to be reliable and accurate.

You need to be able to describe precisely what you did. For example, how many times did you measure something? Was there anything particular about how you measured it? The idea is that someone else should be able to go and copy exactly what you did, using your description as instructions.

Read Brooke's account below of how she measured river width and depth, and try to pick out five things that make it really precise.

> **Revision activity**
>
> Write practice paragraphs about your methods. Make them as precise as possible. Make sure you include details like the ones you identified in Brooke's account.

'We chose places in the river that were safe and accessible. One person walked across the river and held the tape measure at the water surface on the riverbank. The other end was held at the water surface on this side, and the tape measure was pulled tight to make sure it wasn't dangling in the water. We then divided the width by six, and at regular intervals across the river we used a metre stick to measure the depth. We held the metre stick sideways on to the water flow, so that the water could flow past without building up against the metre stick. We made sure the metre stick was touching the river bed, then read where the water level came to.'

> **Exam tip**
>
> You should also be able to write about your equipment. You need to give the name of what you used, and explain how it helped you to be accurate. For example, 'we used callipers to measure stones, as they allowed us to identify the longest part of the stone and hold it steady while we measured. The arrow on the callipers pointed precisely at the correct measurement, making it easier to use millimetres rather than just the nearest centimetre.'

111

Recording sheets

REVISED

These have to be designed carefully to allow you to record the information you need efficiently and clearly. It is also important to write clearly and neatly, so that you can understand your writing later on.

Exam tip

If the exam asks you about recording sheets, you need to be able to talk about your own recording sheet. What did you have spaces for? Did you have a space for a tally chart? Did you record information at several locations separately? Was there space for notes to jog your memory later on?

Secondary sources

REVISED

A secondary source is one where other people have already obtained the information.

Revision activity

Write a few sentences about one secondary source you used. Make sure you can be precise – if it was a map, what sort? What scale? If it was NINIS, which data did you use? How did your secondary source help?

Exam tip

Try to be as precise as possible about a secondary source. For example, 'we used an Ordnance Survey geology map of the area to help us identify the rock types at each site along the river, so that we could understand how they might react to river erosion'.

Exam practice

1 Describe one data collection method **you** used and explain how it helped in **your** investigation. [4]
2 Name one piece of equipment **you** used and explain why it was appropriate for **your** investigation. [3]
3 Describe how **your** recording sheet helped you with your data collection. [3]

Processing and presenting data

> **You need to be able to:**
> - select and use appropriate graphical and cartographic methods, both hand-drawn and using ICT, to process and present your fieldwork data
> - explain why you chose these processing and presentation methods.

Processing data

REVISED

Processing data is what you do when you get back from carrying out your fieldwork. This might mean counting responses to a questionnaire to get a total, or working out average depths in a river. You could do this by hand, to fill in a table, or in a spreadsheet, filling in your data and getting the spreadsheet to give you totals.

> **Revision activity**
>
> Write down on a revision card what **you** did with **your** data when you came back from fieldwork to get it ready to present.

Presenting data

REVISED

Data presentation is where you use graphs and maps to present your data. You have to be able to do this by hand or using ICT. In the exam you will have to draw a graph by hand, but they could also ask how you would do it using ICT.

In the exam they will ask you to draw a graph of some of your data. You will use data on your table. They want to see an appropriate type of graph, drawn accurately. They will check your data table to make sure you were accurate! You will need to be precise, so make sure your pencil is sharp and plot your points as carefully as possible.

You need a title, and it should include reference to both the variables on your graph. For example, you would only get one mark for a title that said 'A graph to show how river width changes', but if you added 'with distance downstream' you would get a second mark. Make sure you label your axes carefully, including the units of measurement.

> **Exam tip**
>
> Make sure your graph is appropriate to your data: for example, bar graph, line graph, scatter graph. For a bar graph, you must have discrete, or separate, data. For a line graph, your data must be continuous – there should be time or distance measurements to go along the x-axis. For a scatter graph, you should have two variables you need to put together. This could include distance as one variable. Make sure you can draw your graph quickly and accurately in an exam.

> **Exam tip**
>
> You need to be able to explain why you chose this type of graph – for example:
> - I chose a scattergraph as it allows me to show the relationship between bedload size and distance downstream. I added a best-fit line as it allowed me to see the relationship more clearly.
> - I chose a line graph because my data is continuous, measuring the number of people in a town centre over one day, so a line graph allowed me to see clearly how this changed during the day. It allowed me to plot separate lines for different days of the week, so I could compare them.
> - I chose a divided bar graph, because it allowed me to clearly show the percentages of bedload stones in each category of roundness. It is easier to compare the figures at different points along the river by placing the divided bars next to each other, rather than drawing separate pie charts for each one.

> **Exam tip**
>
> Practise drawing your graph in advance. You need to know what type, which variable is going on which axis, and what your scale is going to be, using a past paper to check the size of the graph paper you will be given. It is likely to be 1 cm squares, with 2 mm rulings, 18 cm across and 20 cm high. It should be a simple graph – don't be tempted to be too complicated, as you will have to draw it in about 8 minutes, including all the labels! Make sure you have a ruler and a sharp pencil in the exam with you.

113

Now test yourself

1 What variables are you putting on your graph?

2 Which hypothesis does this help with?

3 What is your title?

4 What type of graph are you going to draw? Explain why this is a good graph for your data.

5 Which variable goes on the *x*-axis (across the bottom) and what label are you going to give it? What unit is it measured in?

6 Which variable goes on the *y*-axis (up the side) and what label are you going to give it? What unit is it measured in?

7 What scale are you going to use for each axis?

8 What will you use to help plot your points accurately?

Revision activity

Draw your graph several times. Time yourself to make sure you can do it in 8 minutes!

They could also ask you to describe other methods of presenting your data. This could be about what kind of **map** you could produce, or about how you could use **ICT** to help you present your data.

Revision activity

Have an answer ready for each of these – make sure it is specifically about **your** data. For example, if you are writing about producing a map mention the name of the place, and which data you would display on the map and how. Would it be a piechart on the right locations on the map? Would it be a choropleth map? If it is about using ICT, mention the type of software you might use (such as a spreadsheet, GIS, Google Earth) and be clear which data you would present and how.

Exam practice

1 Choose one hypothesis. Present the relevant data from your table using graph paper. [8]

2 Explain how you could use ICT to produce a graph for one variable in your data table. [3]

3 Describe a map you could produce to display your data. [3]

Exam tip

Remember:
● appropriate graph
● full title
● accurate and precise plotting
● labelled axes including units
● key.

Take a ruler and a pencil into the exam!

Exam tip

Read ahead in the exam. They might ask you to write about one hypothesis early in the exam, then draw a graph for another hypothesis later on. This means you need to spot this at the start, so that you can draw a graph for the data you want, by selecting your other hypothesis for the earlier question, and saving up the one you want for the graph question. If you use the same hypothesis for both sections, you will only be able to get to level 1 in the marks – for example, you might only be able to score 1 mark out of 5.

Analysing and interpreting data

You need to be able to:

- analyse and interpret your fieldwork data using your knowledge of relevant theory and/or case studies
- establish links between data sets
- identify anomalies in your fieldwork data.

Analyse your fieldwork data

REVISED

Data analysis means describing what you can see in your graph. There are four things to do:

- Describe an overall **pattern** or trend. For example, does a variable increase as you go downstream? Are younger people more likely to recycle than older people? Finding links between your two variables is important if possible.
- Give some **figures** to prove your pattern. This might be a figure at the start and finish, or two figures to compare. Try to process the figures if possible – perhaps a figure doubles, or halves, over time, or one figure might be three times as high as another.
- Describe any anomaly – data which does not fit the pattern. You should give a figure for this, and say whether it is higher or lower than you would expect.
- Say whether it supports your hypothesis.

Revision activity

Read the analysis below. Use four colours to highlight the pattern, figures, anomaly and hypothesis:

'The graph in Figure 1 shows that cross-sectional area increases as you go further downstream. For example, at 0.3 km downstream the cross-sectional area is 0.09 m², whereas at 9.1 km downstream the cross-sectional area has increased to 2.84 m². There is one significant anomaly at 6.2 km downstream, where cross-sectional area is higher than we would expect, at 3.3 m². This graph supports the hypothesis that cross-sectional area increases as you go downstream.'

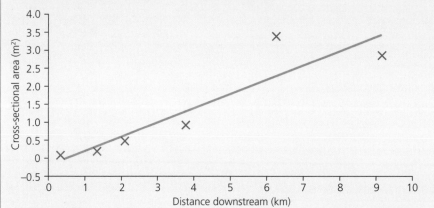

Figure 1 A graph to show how cross-sectional area changes with distance downstream.

Revision activity

Practise writing about what **your** graph shows. Use three colours to make sure you include a pattern, figures and any anomalies. Then use a fourth colour to state what it shows about **your** hypothesis – does it support your hypothesis or not? You should do this several times so that it is easy in the exam.

Exam tip

In the exam you will have your table of data, so don't waste time trying to read figures off your graph, just use the table to help!

Now test yourself

TESTED

1 What are the four things you need to remember when you are analysing your data?
2 What do we call data that doesn't fit the pattern?

Interpret your fieldwork data

Interpretation is where you have to explain what your graph shows, using your knowledge of geographical theory. You should explain the pattern using theory. You must also explain why any anomalies exist – this will probably need specific information from your fieldwork. For example, if you have counted fewer people than you expected somewhere, is it because a road was closed? Or because it was raining when you counted in that area?

Exam tip

Remember, your graph might agree with the theory, or it might not! That's OK so long as you can explain why it might be different from the theory!

Revision activity

Read the explanation below, and use a colour to pick out where theory has been used, and another to show where the anomaly has been explained.

'The cross-sectional area of the river increases as you go downstream because tributaries flow into the river, and add more water to it. This means the channel has to become wider or deeper to hold the additional water. The river erodes downwards from the river bed, and sideways at the banks by processes such as abrasion. This is where the stones it is carrying along rub against the rock and wear it away, which makes the river channel deeper and wider. The anomalous result at 6 km is probably because it is just after a major confluence which added water from several tributaries. This increases the volume of water and its velocity, so it can erode the river bed and banks more quickly, making the channel deeper and wider.'

Exam tip

An exam question that asks you to suggest a geographical reason for the pattern or relationship shown in your graph is asking you to use your theory to explain it, and your specific knowledge about the location to explain any anomalies.

Revision activity

Now practise giving an explanation of **your** data. Use two colours to make sure you use theory, and explain any anomalies.

Exam practice

1 Describe what your graph shows about your hypothesis.   [4]
2 Explain what your graph shows, using appropriate geographical theory.   [6]

Exam tip

In question 1, remember: pattern, figures, anomalies, hypothesis.

Drawing conclusions

You need to be able to:

● draw conclusions about your hypotheses, and back them up with evidence.

A conclusion is very simple – you state whether each hypothesis was supported, not supported, or perhaps partly supported. Then you back up what you have said with evidence from your graph and explanations. You should also explain how this helps with the aim of your study.

Revision activity

Read the conclusion below. Use four colours to pick out the hypothesis, whether it is supported, the evidence used and how it helps with the aim.

'My hypothesis was that cross-sectional area increases as you go downstream. This is supported by my results. Cross-sectional area increased from 0.9 to 2.84 m², and this is due to increased water volume, velocity and erosion processes such as abrasion. This helps with the aim of the study, as it helps to confirm that part of the Bradshaw model was correct for this river.'

Revision activity

Write a conclusion for each of your hypotheses, based on the evidence you are using for your graph. Use four colours to make sure you include the hypothesis, whether it is supported, some evidence to prove it, and how it helps with the aim of your study.

Exam practice

What conclusions can you draw from your data in relation to your chosen hypothesis?   [4]

Exam tip

Don't forget to use data to back up what you say.

Evaluating your fieldwork

You need to be able to:

● describe your data collection methods, including any equipment used;
● identify problems with your data collection methods;
● identify limitations of the data collected;
● suggest other data that might be useful;
● evaluate your conclusions; and
● suggest how you could extend the scope of the study.

Describing your data collection methods   REVISED ○

We covered this earlier. Remember it is important to be properly precise about how **you** collected **your** data.

Problems with data collection methods   REVISED ○

Think about when you were doing your fieldwork. What methods gave you problems?

For example, when Niamh was carrying out a survey in the town centre, she found lots of people didn't want to stop and answer her questions. Some older people were happy to, but younger people seemed too busy. She was concerned that this might bias her data towards older people.

Zeke visited the river one day after heavy rain. This meant it was flowing very high, so he couldn't get to all the places he wanted to measure as it wasn't safe. He had to stick to shallower slow-flowing places, so his data did not represent the deeper faster areas.

Revision activity

Write a few sentences about **your** data collection methods. What gave **you** trouble?

117

Limitations of the data

REVISED

This is about the amount of data you have, how reliable it is and how clear it is.

Issue	Meaning	Examples
Amount	Do you have enough data to represent a larger place or group?	Did you ask enough people in your survey to be representative of a larger group? Did you measure the river at enough places to give a good idea of change downstream? Did you take several measurements and use an average, or did you only have one measurement?
Reliability	Was anything unusual happening that could have affected your data?	Was there an event in the news that might have changed people's answers to a questionnaire? Was there a rainstorm or a drought that could affect river flow?
Clarity	Are the results clear?	Did everyone follow the instructions correctly or could there be human error? Did everyone work accurately, or were some people careless?

Revision activity

Use the example below to help **you** identify limitations with **your** own data. Make sure you explain them fully:

'We measured the river at six locations downstream, but there were large gaps between measurements in some places, so it might have been better to have more locations. The river was flowing quite high on our fieldwork day, so cross-sectional area was higher than usual, and large rocks may have been carried further downstream than usual because of the higher river velocity. Measuring the velocity was very difficult as the water was turbulent, so the piece of wood that we floated downstream and timed kept getting stuck.'

Now test yourself

List the three ways in which data can have limitations.

TESTED

Suggest other data that might be useful

REVISED

When you wrote **your** interpretation, is there anything that would have helped?

Revision activity

Think about:
● **Primary data** – is there anything else **you** could have measured? Could you have taken photographs – would they have helped?
● Secondary data – is there anything **you** would have liked to know about the area? Would it help to know about rock type? Land use?

Evaluate your conclusions

REVISED

This means deciding on how reliable your conclusions are. If there were big problems with your data collection methods, or significant limitations with amount, reliability or clarity of your data, then the conclusions are not totally reliable – they can be questioned.

Revision activity

Look over **your** answers for the last few sections. What do you think overall? Are **your** conclusions reliable? Or could they be questioned? Why?

Exam tip

You won't lose marks for pointing out the weaknesses in your fieldwork. It will actually gain you marks if you can take a step back and recognise its flaws.

Suggest how you could extend the scope of your study

This means, if you had more time, how could you make this bigger? For example, can you compare your study to somewhere else? Could you go back in a year's time and see if anything has changed? Could you compare winter and summer? Could you add another hypothesis? How would this help – it might enable you to find out more about your aim overall? Can you be specific?

For example:
'Our aim was to investigate how river characteristics change along the Curly Burn. We could have extended the study by measuring other variables, such as the gradient of the river. This would allow us to investigate other relationships such as the relationship between velocity and gradient, and therefore to find out more about how the Curly Burn changes downstream.'

> **Revision activity**
>
> Think about this for **your** study – what could **you** do?

> **Exam practice**
>
> 1 Describe one data collection method you used, and explain any problems you encountered with this. [5]
> 2 Evaluate your data, identifying any limitations. [3]
> 3 Identify any additional information that would have been useful to you in your investigation and explain how it would have helped. [3]
> 4 Evaluate your conclusions. What could you do to make them more reliable? [4]
> 5 Give two ways you could extend the scope of your study. [4]

> **Exam tip**
>
> This paper is a gift – it is easy to practise every section, and you are writing about fieldwork, which is probably quite memorable! You just need to make sure everything you write is very specific, so that someone who wasn't on your fieldwork gets a really good idea of what you did, where, why, how, and what your results show.

> **Exam tip**
>
> Some of these questions will seem repetitive – you wouldn't be asked all of these, but you should have an answer ready for any of them.

Glossary

abrasion/corrasion (1) The grinding of rock fragments carried by a river against the bed and banks of the river. (2) A process of erosion which occurs when a wave hits the coast and throws pebbles against the cliff face. These knock off small parts of the cliff, causing undercutting. Page 17

afforestation the planting of trees in an area that has not been forested before. Page 22

aged dependent the proportion of people aged 65 or over in a population, compared to those of working age (15–64). Page 66

air mass a body of air with similar characteristics, for example, temperature, humidity and air pressure. Page 44

altitude height above sea level, normally given in metres. Page 42

anemometer an instrument which is used to measure wind speed. It has mounted cups which revolve faster or slower depending on the wind speed. Page 39

anomaly a result which is unexpected. Page 115

anticyclone a weather system with high pressure at its centre, generally associated with dry, calm weather. Page 45

appropriate technology technology which is suited to the level of development in the area where it is used. Page 90

arch a wave-eroded passage through a small headland. This begins as a cave formed in the headland, which is gradually widened and deepened until it cuts through. Page 29

atmospheric pressure the weight of a column of air measured in millibars. Page 39

attrition a process of erosion where transported particles hit against each other, making the particles smaller and more rounded. Page 17

barometer an instrument used to measure air pressure. Page 38

beaches form in sheltered coastal areas, where the sea deposits sediment such as sand or stones (shingle) along the coastline. A sandy beach usually has a gentle slope up to the land, while a shingle beach is usually steeper. Page 30

beach nourishment the addition of new material to a beach artificially, through the dumping of large amounts of sand or shingle. Page 34

biofuel a fuel that comes from living matter, such as plant material. Page 99

BRICS Brazil, Russia, India, China, South Africa. Large and growing economies that contribute to global patterns of trade and interdependence. Page 94

buoy an anchored float which may contain instruments to measure environmental conditions. Page 39

carbon footprint a measure of the amount of carbon dioxide produced by a person, organisation or country in a given time. Page 96

caves hollows in a cliff or headland made when the sea erodes the rock around a line of weakness. Page 29

central business district (CBD) the part of the settlement which is dominated by shops and offices, often has tall buildings and is usually close to its centre. Page 74

cirrus a type of cloud, it appears as high, wispy and white and is made from ice crystals. Page 39

cliff a near-vertical rock face at the coast. Page 29

climate the average weather conditions of an area over a long period of time, for example, 35 years, covering only temperature and precipitation. Page 37

climate change changes in long-term temperature and precipitation patterns which increasing evidence suggests is caused by human activities. Page 96

cloud cover the amount of sky covered by cloud, measured in oktas. Page 39

cloud types clouds are divided into categories (nimbus, cirrus, cirrocumulus and so on) depending on their height, shape and nature. Page 39

coastal defences natural or manufactured management strategies which try to maintain the land–sea boundary or reduce the impact of erosion. Page 32

coastal landforms large features within a coastal area formed by processes such as erosion and deposition. Page 30

coastal management strategies a variety of approaches taken to control the interaction between people and the coast. Page 33

cold front the zone where cold air comes behind warm air. The cold air undercuts the warm air, forcing it to rise, cool and condense. Page 45

cold sector the larger part of a depression which contains polar maritime air. Page 45

collision zone plate margin where two continental plates move towards each other and create fold mountains. Page 53

composite volcano a steep-sided, dome-shaped volcano that erupts a variety of materials such as sticky acidic lava and ash. Occurs at destructive plate margins. Page 61

conclusion the findings of a study summarised. Page 117

confluence the point where two rivers meet. Page 13

conservative margin when plates slide past each other along a fault line, and crust is neither destroyed nor added to. Page 54

constructive wave a wave with a strong swash and weak backwash which contributes deposition to a beach. Page 25

convection current repetitive movements set up in the mantle due to heating by the core. These currents make the crust move. Page 52

core the centre of the Earth, found below the mantle. It is extremely hot and may be made of metal. Page 52

corrasion an alternative word for abrasion. Page 17

corrosion the process by which water (in river or sea) reacts chemically with soluble minerals in the rocks and dissolves them. Page 17

crude birth rate total number of live births per thousand of the population per year. Page 63

crude death rate total number of deaths per thousand of the population per year. Page 63

crust the upper layer of the Earth on which we live. It is solid but is split into sections called plates. Page 52

cumulonimbus a dense, tall, towering cloud which often has a flat, anvil top. They may produce heavy rain and thunderstorm conditions. Page 39

cumulus a common cloud type with puffy white tops and often a flat base. They are a low to mid-level cloud. Page 39

dam a barrier (made of earth, concrete or stone) built across a valley to interrupt river flow and create a man-made lake (reservoir) which stores water and controls the discharge of the river. Page 22

data analysis a written, detailed, examination of a graph, map or table to establish trends and anomalies in the data gathered. Figures are quoted and simple calculations carried out. Page 115

data collection the process of gathering primary information from the field, for example, questionnaire, traffic counts, river velocity recording. Page 111

data presentation how data is displayed to readers: usually in words, graphs or tables. Geographical data collected should be presented as tables, annotated photographs, maps or graphs. Page 113

demographic transition model a theoretical model based on the experience in MEDCs showing changes in population characteristics over time. Page 64

dependency a measurement of the number of people of working age compared to the children and elderly people who need to be supported by them. Aged dependency means that there is a large group of elderly people (65+) and youth dependency means there is a large group of children and young people (under 15). Page 66

deposition the dropping of material on the Earth's surface. Page 16

depression a weather system with low pressure at its centre, characterised by wet and windy conditions. Page 45

depth (river) the distance from the surface of a river to the river bed. Page 15

destination countries countries to which migrants move. Page 71

destructive margin where one plate crashes into another plate and crust is destroyed. Page 57

destructive wave a wave with a strong backwash and weak swash which erodes a coast. Page 25

development the level of economic growth and wealth of a country. The use of resources, natural and human, to achieve higher standards of living. This can include economic factors, social measures and issues such as freedom. Page 85

development gap the division between wealthy and poor areas, in particular the disparity between LEDCs and MEDCs. Page 85

digital thermometer an instrument that measures the temperature of the air, displaying the result as an LCD readout. Page 39

discharge the amount of water in a river which is passing a certain point in a certain time. It is measured in cumecs (cubic metres per second). Page 15

drainage basin an area of land drained by a river and all of its tributaries. Page 12

earthquake a tremor starting in the crust which causes shaking to be felt on the Earth's surface. Page 53

economic indicators figures relating to the wealth and economy of a country. Page 86

economic migrant someone who moves because they want to improve their chance of getting employment and earning money. Page 71

ecotourism otherwise known as green tourism. A sustainable form of tourism which involves protecting the environment at the destination. Page 104

embankments artificial mounds along a river bank to prevent a river bursting its banks. Page 22

emigration the movement of people away from a country. Page 70

epicentre the first place on the Earth's surface to feel shockwaves from an earthquake. It is directly above the focus. Page 58

erosion wearing away of the landscape by the action of ice, water and wind. Page 16

evapotranspiration the process by which water is transferred from the land to the atmosphere by evaporation from surfaces, for example, lakes, and by transpiration from plants. Page 12

fair trade a type of trade where producers in a poor country get a fair living wage for their product and which promotes environmental protection. Page 92

fault line a weak line in the Earth's surface, where the crust is moving, causing earthquake activity. Page 53

flood wall a stone/brick/cement wall built alongside a river to protect the nearby areas from flooding. Page 22

flooding a temporary covering by water of land which is normally dry. Page 21

floodplain flat land either side of a river, which will be flooded if a river bursts its banks. Page 18

focus the point of origin of an earthquake under the Earth's surface. Page 58

fold mountains mountain ranges that form mainly by the effects of folding of the Earth's crust at destructive and collision plate margins. Page 53

front the zone where two types of air mass meet. Page 45

frontal depression a weather system with low pressure at its centre and two contrasting air masses. They are associated with wet and windy weather. Page 45

function role performed by something, such as a city. A function of one part of the settlement might be to provide housing, or employment. Page 75

gabions metal cages filled with rocks which can form part of a sea defence structure or be placed along rivers to protect banks from erosion; an example of hard engineering. Page 33

gentrification the renovation of homes and businesses in run-down inner-city areas so that they appeal to better-off people, but original communities are pushed out by the higher cost of houses. Page 74

geostationary satellite a satellite that is positioned over one place and moves at the same speed as the Earth. It only provides data for that one place. Page 39

global warming the warming of the atmosphere; that is, the increase over time in average annual global temperature. This is probably related to human activity through the release of greenhouse gases. Page 32

globalisation the way in which countries from all over the world are becoming linked by trade, ideas and technology. Page 93

gradient the steepness of the slope. Page 15

greenhouse effect a natural process where our atmosphere traps heat. Some of the heat from the Sun that is absorbed by the Earth's surface is re-radiated to the atmosphere where it is held by the greenhouse gases – carbon dioxide, methane, nitrogen dioxide, CFCs and water vapour. Page 95

groundwater water stored in the bedrock. Page 98

groundwater flow water which is moving horizontally through the bedrock towards a river or sea. Page 12

groynes wooden barriers built out into the sea to stop the longshore drift of sand and shingle, and so cause the beach to grow. They are used to build beaches to protect against cliff erosion and provide an important tourist amenity. However, by trapping sediment

they deprive another area, down-drift, of new beach material. Page 33

hard engineering methods strategies to control a natural hazard which do not blend into the environment. Page 22

headland an area of land that extends out into the sea, usually higher than the surrounding land; also called a point. Page 29

hooked spit a coastal landform caused by deposition and the transport process of longshore drift. Changes in the prevailing wind and wave direction can cause a spit to form a curved, hooked end. Page 30

Human Development Index (HDI) a measure of development which combines measures of wealth, health and education, thus mixing social and economic indicators. Page 87

hydraulic action (1) A form of erosion caused by the force of moving water. It undercuts riverbanks on the outside of meanders and forces air into cracks in exposed rocks in waterfalls. (2) The process whereby soft rocks are washed away by the sea. Air trapped in cracks by the force of water can widen cracks, causing sections of cliff to break away from the cliff face. Page 17

hypothesis a statement, believed to be true, which is to be tested during an investigation. For example, river load will decrease in size downstream. Page 106

igneous when referring to rocks, this means rocks formed when molten magma either hardens under the crust or erupts through and hardens on the Earth's surface. Page 53

immigration the inward movement of people to one country from another. Page 70

infiltration the movement of water into the soil from the Earth's surface. Page 12

inner city the area surrounding a CBD, characterised by mixed land use, with some older terraced residential or newer apartments mixing with industry. Page 74

inner core the Earth's innermost layer, thought to be solid and made from the metals iron and nickel. Page 52

interception the process whereby precipitation is prevented from falling onto the ground by plants. It slows runoff and reduces the risk of flash flooding. Page 12

interpretation making sense of the data by giving explanations for trends identified in the analysis. These may be related to theory or local geographical causes. Page 116

knot the unit used to measure wind speed. Page 38

land use zones areas of a settlement which share the same function – such as housing, industrial or commercial. Page 22

landform a natural, recognisable feature of the Earth's surface. Page 29

Exam practice answers and quick quizzes at **www.hoddereducation.co.uk/myrevisionnotesdownloads**

latitude the imaginary lines that surround the Earth ranging from 0° at the equator to 90° at the poles. They tell us how far a location is north or south of the equator. Page 41

LEDC a less economically developed country, often recognised by its poverty and a low standard of living. Page 85

levees raised banks along a river that help to reduce the risk of flooding. Page 18

liquefaction the process of solid soil turning to liquid mud caused by shaking during an earthquake bringing water to the surface. Page 58

load the sediment carried by a river. Page 15

longshore drift the process whereby beach material moves along a coastline, caused by waves hitting the coast at an angle. Page 28

managed retreat moving land uses away from the coast. Page 34

mantle the layer above the Earth's core. It makes up 80% of the Earth's mass. It behaves like liquid rock. Page 52

mass tourism large-scale movements of people to tourist destinations which can result in the building of hotels in vulnerable areas and can have a negative impact on the relationship between local communities and the visitors. Page 102

meander a river landform. A sweeping curve or bend in the river's course. Page 18

MEDC a more economically developed country, often recognised by its wealth and a high standard of living. Page 85

metamorphic rocks that have been changed as a result of heat and pressure being applied to them over long periods of time. Page 55

mid-ocean ridge where two plates made of oceanic crust move apart, the magma of the mantle rises to fill the gap, causing the crust to rise and form a ridge. Page 53

migration the permanent or semi-permanent movement of people from one place of residence to another. Migration can be classified, for example, into forced, such as due to war or famine, or voluntary, such as looking for better work. Page 70

millibar the unit used to measure air pressure. Page 38

natural change the difference between the number of deaths and the number of births in a place. Page 63

natural decrease when the death rate is higher than the birth rate, there is a natural decrease in population. The population will fall, unless there is more immigration than natural decrease. Page 63

natural increase when the birth rate is higher than the death rate, there is a natural increase in population. The population will rise, unless there is more emigration than natural increase. Page 63

ocean trench a feature of a destructive plate margin which involves oceanic crust. Where the oceanic crust is forced down into the mantle it sinks below its normal level to create a deep trench in the ocean floor. Page 53

okta a measure of cloud cover, estimated as eighths of the sky covered in cloud. Page 38

outer core a layer of the Earth found below the mantle, made from molten metals. Page 52

percolation the movement of water from the soil into the bedrock. Page 12

plate margin a zone where two plates meet. Plate margins may be described as constructive, destructive, conservative or collision. Page 53

polar continental air mass a large body of air coming from the east over the British Isles. Page 44

polar maritime air mass a large body of air coming from the north-west over the British Isles. Page 44

polar satellites instruments based in space that orbit the Earth as they record information, making orbits roughly 14.1 times daily. Page 39

population change an increase or decrease in the number of people in a country or area. Page 64

population pyramid a type of bar graph that shows the structure of a population by sex and age category, and may resemble a pyramid shape. Page 65

population structure the composition of a population, divided up by age group or gender. Page 65

precipitation a form of moisture in the atmosphere, such as rainfall, sleet, snow and fog. Page 12

prevailing wind the most frequent, or common, wind direction. Page 41

primary activity an activity which uses the Earth's resources as a way of making money, for example, farming, fishing, agriculture and mining. Page 88

primary data data collected by students personally during fieldwork as a result of measurement and observations. Page 118

primary sources elements in the environment which can be measured using observation and equipment, such as cloud cover or pedestrian numbers. Page 109

pull factor any attractive/positive aspect or quality of a place which attracts (pulls) migrants to it. Page 70

push factor any negative aspect or quality of a place which causes people to leave it. Page 70

rain gauge an instrument which catches and measures precipitation. Page 39

rainfall radar also known as a Doppler weather radar, this is a type of radar used to locate precipitation, predict its motion and estimate its type (rain, snow, hail). Page 39

refugee a person who has been forced to leave their home country and move to another country, often in response to persecution or natural disaster, and has applied for refugee status in the destination country. Page 71 regeneration the improvement of part of a settlement through rebuilding, restoration or the removal of pollution. Page 78

renewable energy source a sustainable source of electricity production such as wind, solar or biofuels. Page 99

responsible tourist a tourist who respects the environment and the people of the places which he or she visits. Page 104

Richter scale a scale between 0 and 9 which measures the strength of an earthquake. Page 58

risk a judgement of the potential for coming to harm in a given situation. Page 108

river cliff a small cliff found on the outside bend of a meander, formed by the river eroding the bank. Page 19

river landforms large-scale features found along the course of a river, such as waterfalls and meanders. Page 18

river mouth the end of a river where it meets the sea, ocean or lake. Page 13

rock types rocks are divided into categories depending on how they were formed: igneous, sedimentary and metamorphic. Page 55

rural–urban fringe an area on the outskirts of a city beyond the suburbs where there is a mixture of rural and urban land uses. Page 74

saltation the transportation of medium-sized particles by bouncing them along. Page 17

sandy beach a coastal landform, caused by deposition, which mostly consists of very small mineral particles and has a shallow gradient. Page 30

satellite image a photograph or remotely sensed image recorded from space. Page 48

sea walls curved concrete structures placed along a sea front, often in urban areas such as the front of a promenade, designed to reflect back wave energy; an example of hard engineering. Page 33

secondary data data collected from sources other than the student; may include published material, reports from public bodies and the work of other people. Page 106

secondary source source of data collected by others rather than yourself (primary data). Page 109

sedimentary rocks that have been produced from layers of sediment, usually at the bottom of the sea. Page 55

seismograph an instrument designed to measure the energy released by earthquakes. Page 58

settlement a place where people live and which provides services and places of employment and entertainment. Page 75

shanty town a characteristic of LEDC cities; an area of unplanned poor-quality housing which lacks basic services like clean water. Page 83

shield volcano a wide, low volcano that erupts basic runny lava. Occurs at constructive plate margins. Page 61

shingle beach a coastal landform, caused by deposition, which mostly consists of medium-sized mineral particles and has a steep gradient. Page 30

slip-off slope a small beach seen on the inside of some meander bends which is caused by deposition in shallow water. Page 19

social indicators pointers, usually of level of development, which are to do with people. Examples include quality of health and education. Page 86

soft engineering methods strategies to control a natural hazard which blend into the environment, so are often sustainable. Page 22

solar energy the Sun's energy exploited by solar panels, collectors or cells to heat water or air or to generate electricity. Page 99

solution the process by which water (in river or sea) reacts chemically with soluble minerals in the rocks and dissolves them. Page 17

source the starting point of a river, it may be a lake, glacier or marsh. Page 13

spit a depositional landform formed when a finger of sediment extends from the shore out to sea, often at a river mouth. Caused by the transport process of longshore drift. Page 30

stacks natural features of an eroded cliff landscape that appear like large free-standing sections of coastline. Stacks are formed by the collapse of a sea arch. Page 29

stratus a type of cloud that appears as a continuous flat sheet of grey cloud. Page 39

subduction the sinking of a dense plate into the mantle. Page 53

subduction zone an area where crust is being forced down into the mantle. Page 53

suburb a residential area or a mixed use area usually at the edge of a settlement, but within commuting distance of the centre of a city. Page 74

supervolcano a type of volcano where the potential exists of it erupting at least 1000 km3 of material, having global consequences. Page 61

surface runoff/overland flow water which is moving over the surface of the land. Page 12

suspension the transportation of the smallest load, for example, fine sand and clay which is held up continually within river water or seawater. Page 17

Sustainable Development Goals seventeen global goals with 169 targets adopted by countries around the world in 2015. Spearheaded by the United Nations, these emphasise improving current quality of life while still maintaining resources for the future. Page 89

sustainable tourism tourism which does not damage the place where it happens, allowing it to continue to be used by future generations. Page 102

Exam practice answers and quick quizzes at **www.hoddereducation.co.uk/myrevisionnotesdownloads**

synoptic chart a weather map which shows the weather as symbols over an area. Page 47

tectonic plates the crust is broken up into seven large sections and various smaller sections, which are floating on the mantle and moving towards, away from and past each other. Page 52

temperature how hot or cold the air is. Usually measured in degrees Celsius. Page 39

throughflow water which is moving through the soil. Page 12

traction the rolling of boulders and pebbles along the river bed. Page 17

transportation the movement of material across the Earth's surface. Page 16

tributary a stream which flows into a larger river. Page 13

tropical continental air mass a large body of air coming from the south-east over the British Isles. Page 44

tropical maritime air mass a large body of air coming from the south-west over the British Isles. Page 44

tsunami a large wave caused by an underwater earthquake. Page 58

urban planning schemes plans, usually on a large scale, for changing parts of a settlement. Sometimes these change the use of an area from a former industrial area to housing or commercial uses. Page 82

volcano a cone-shaped mountain built up from hardened ash and lava, from which molten material erupts onto the Earth's surface. Page 53

warm front the zone where warm air arrives behind cold air in a depression. Page 45

warm sector the smaller part of a depression which contains tropical maritime air. Page 45

washlands an area deliberately left to accommodate flood water from a river. This could be farmland or recreational land close to settlements. Page 22

waste hierarchy the arrangement of waste disposal options in order of sustainability. Page 98

water cycle the continuous circulation of water between land, sea and air. Page 12

waterfall a steep fall of river water where its course crosses between different rock types, resulting in different rates of erosion. Page 18

watershed the boundary between drainage basins, it is often a ridge of high land. Page 13

wave cut platform a flat area along the base of a cliff produced by the retreat of the cliff as a result of erosive processes. Page 29

weather the day-to-day condition of the atmosphere. The main elements of weather include rainfall, temperature, wind speed and direction, cloud type and cover, and air pressure. Page 37

weather forecast a description of the state of the weather in an area with projections on how it might also change in the next few days. Page 39

width (river) the measurement from one river bank to the other across a river channel. Page 15

wind direction the geographical direction (compass point) from which a wind blows. Page 39

wind energy a form of renewable energy, where wind turbines convert the kinetic energy in the wind into mechanical power and eventually into electricity. Page 99

wind speed the speed at which air is flowing. It can be measured in knots. Page 39

wind vane an instrument used to measure wind direction using a fixed directional compass and a movable pointer. Page 39

youth dependent the proportion of people aged fourteen or under in a population, compared to those of working age (15–64). Page 66

Now test yourself answers

Page 10

1 Flat land
2 59 m
3 Residential, car park, caravan park/tourism
4 SE
5 Floodplain and meander
6 Output
7 Suburbs
8 Bus station, cathedral, museum, inside the ring road.

Page 13 (1)

1 A Groundwater.
   B Interception.
   C Precipitation.
   D Throughflow.
   E Infiltration.
   F Evapotranspiration.
   G Groundwater flow.
   H River discharge.
2 Precipitation.
3 Interception.
4 Groundwater flow.
5 Water moving downwards from soil into bedrock.
6 Output.
7 Evaporation.
8 Infiltration goes downwards into the soil. Throughflow flows through the soil into the river.

Page 13 (2)

   A – 2
   B – 4
   C – 5
   D – 3
   E – 1
   F – 6

Page 15

1 Gradient: steep/gentle.
   Width: narrow/wide.
   Depth: shallow/deep.
   Discharge: low/high.
   Load: large, angular/small, rounded.
2 Tributary.
3 Confluence.
4 The high land around the edge of a drainage basin.

Page 16

1 (a) Hydraulic action.
   (b) Suspension.
   (c) Erosion.
   (d) Deposition.
   (e) Saltation.
   (f) Attrition.
2 They crash into each other and the river bed and banks, which breaks them up and smooths the rough edges, so they become smaller and rounder.
3 Abrasion.
4 Saltation.
5 The river has less energy, so it drops its load – this is known as deposition.

Page 18

1 Undercuts.
2 The overhanging hard rock falls into the plunge pool.
3 Hydraulic action and abrasion make a plunge pool.
4 Waterfall moves backwards.
5 Soft rock.
6 Hard rock.

Page 19

1 Water flows faster around the outside bend of the meander. It has lots of energy, so it erodes the bank, moving it back, and erodes the bed, making it deeper. On the inside of the meander, water flows more slowly, so it has less energy and deposits its load, making shallow water.
2 At C, fast-flowing water, deep water, erosion, river cliff. At D, slow-flowing water, shallow water, deposition, slip-off slope.
3 This task requires a sketch of any of the meanders on the river. Outside bends should be labelled with erosion, and inside bends with deposition.
4 Deposition takes place on the inside bends, where there is yellow visible.

Page 20

1 A Bluff.
   B Floodplain.
   C River channel.
   D Levee.
2 Plunge pool.
3 True – it erodes on the outside bank because the water is travelling fastest there. However, it is deep because of the erosion.
4 Sediment/alluvium.

Page 23

1 Dams, levees, straightening the river.
2 Afforestation, safe flooding zones.
3 15 m.
4 Over 100.
5 Rock Island, Illinois.
6 1750 km.
7 4.3 m.
8 Heron.

Page 26

1 A Higher wave.
  B Waves close together.
  C Strong backwash pulls sediment down the beach.
  D Waves far apart.
  E Lower wave.
  F Strong swash pushes sediment up beach.
  G Water sinks into beach, reducing backwash.
2 Destructive.

Page 27

1 (a) Abrasion.
  (b) Attrition.
  (c) Hydraulic action.
2 1 – b
  2 – d
  3 – a
  4 – c

Page 28

Longshore drift.

Page 29

1 1 – d
  2 – a
  3 – c
  4 – b
2 1 – c
  2 – a
  3 – d
  4 – b

Page 30 (1)

1 1 – d
  2 – b
  3 – c
  4 – e
  5 – a
2 Spit is the odd one out because it is formed by deposition, whereas the others are all formed by erosion.

Page 30 (2)

1 a – 3
  b – 4
  c – 2
  d – 1
2 Sediment must be moving from north to south.

Page 32

Global warming will cause sea level to rise by 0.5–1 m. This means that more places will be affected by the sea, and storms will have greater effects in terms of flooding and erosion. This means more coastal defences may be needed.

Page 34

1 (a) Sea walls may be made with steps to break up the sea's energy.
  (b) Sea walls may be curved to reflect wave energy back out to sea.
2 Beach nourishment.
3 Groynes.

Page 36

1 Groynes and gabions.
2 This is a matter of opinion.
  Groynes – not sustainable as they have to be replaced.
  Gabions – sustainable as they are effective.
  Sea wall – unsustainable.

Page 37

(a) Weather.
(b) Weather.
(c) Climate.
(d) Climate.
(e) Weather.
(f) Climate.

Page 38

Temperature – digital thermometer – degrees Celsius.
Precipitation – rain gauge – millimetres.
Wind speed – anemometer – knots.
Wind direction – wind vane – eight points of the compass.
Cloud type – observation – oktas.

Page 40

1 Anemometer.
2 Rain gauge.
3 Millibars.
4 Degrees Celsius.
5 Cloud cover.
6 Stevenson screen.

7 Allows air to flow through but keeps the instruments out of direct sunlight.

8 In the open so there is nothing stopping the rain.

9 Weather buoys, satellites, rainfall radar or land-based weather stations.

Page 43

1 (a) Altitude. A is warmer.
   (b) Distance from the sea. A is warmer in summer, B is warmer in winter.
   (c) Latitude. A is warmer.
   (d) Prevailing winds. B is warmer.

2 Latitude.

3 It is hotter near the equator because the sun is overhead, so its rays are concentrated in a small area which gets heated up.

4 Prevailing wind.

5 Land.

6 Land.

7 Belfast is near the sea. This takes a long time to warm up, so during the summer it is still cool, and keeps Belfast cool. Moscow is a long way from the sea. The land heats up quickly, and it is not cooled down by the sea.

8 Altitude.

9 10°C.

10 Belfast.

Page 44

1 Cool, cold, wet.

2 Hot, cold, dry.

3 Hot, mild, dry.

4 Warm, mild, wet.

Page 46

1 Polar maritime; cool in summer, cold in winter, wet.

2 (a) High pressure. (b) Sinking air.

3 (b) The warm and cold fronts.

4 Depression.

5 Anticyclone.

Page 48 (1)

1 Depression.

2 Cold front.

Page 48 (2)

1 Long range means a long time in advance: 3 months.

2 Short range means a short time in advance: the next 24 hours.

3 Short range is more likely to be accurate.

Page 51

1 134 people were killed.

2 Barbuda.

3 6.5 million.

4 185 mph.

5 The total cost of the damage was $65 bn.

Page 54

1 Your diagram should have the correct shape, and include all the key features from Figure 3a.

2 Convection currents in the mantle drag the plates along.

3 Constructive.

4 Destructive.

5 Constructive.

6 Subduction.

7 Plates move alongside each other.

Page 55

Igneous – extrusive.
Igneous – intrusive.
Sedimentary.
Sedimentary.
Metamorphic.
Metamorphic.

Page 56

1 (a) Extrusive igneous.
   (b) Basalt.
   (c) Intrusive igneous.
   (d) Granite.
   (e) Sedimentary.
   (f) Sandstone/limestone.
   (g) Metamorphic.
   (h) Marble/slate.

2 Sedimentary.

3 Granite.

4 Limestone, sandstone.

5 (a) Slate. (b) Marble.

Page 58 (1)

1 Focus.

2 Epicentre.

Page 58 (2)

1 Liquefaction.

2 Tsunami.

Page 60

1 Pacific, Eurasian.

2 9.1.

Exam practice answers and quick quizzes at **www.hoddereducation.co.uk/myrevisionnotesdownloads**

3 The tsunami meant many were drowned.

4 500,000.

5 Preparation in Japan was effective in some ways, as people get warnings on the TV, interrupting programmes, and on their mobile phones. However, people are so used to getting warnings of earthquakes and tsunamis that many did not take action urgently, and so they were caught by the tsunami, meaning it was not fully effective. Buildings are built to withstand earthquakes, and this was effective, as many were still standing after the earthquake. However, the tsunami was higher than the protective walls, and the powerful fast-moving water destroyed 332,395 buildings. Overall, the preparation was good but not adequate for an earthquake and tsunami this big.

6 The response was effective in some ways. The army acted quickly to build shelters, and 163 countries offered help. Initially some shelters were poorly equipped, although this was improved quickly. The government set up the Reconstruction Design Council which had a budget of over 23 trillion Yen to rebuild houses very quickly. This was effective, as by 2018 house building was 94% complete and 99% of roads were rebuilt. However, this was not fully effective, as some people were still living in temporary accommodation for several years. Overall, the response was highly effective, but still had some problems.

Page 61

1 (a) Composite. (b) Shield. (c) Supervolcano.
2 Shield volcanoes are wider and lower than composite volcanoes.
3 A supervolcano releases about 1000 times the amount in a normal volcanic eruption. They create a big depression called a caldera when the magma chamber underneath is empty.
4 Composite and supervolcano.
5 Shield volcano.

Page 62

1 90,000.
2 Bison and wolves.
3 Global temperatures reduced by 10°C for ten years.
4 Asia.
5 Buildings could be destroyed by the weight of only 30 cm of dry ash in places as far away as Los Angeles and Chicago.

Page 64

1 Food supplies and healthcare were poor.
2 Better food and healthcare.
3 People knew that several of their children were likely to die, so they had lots of children.
4 People realised more of their children were surviving.

5 Stages 2 and 3.
6 More of the population are older, so therefore more of the population are likely to die of old age.

Page 65

1 False.
2 True.
3 True.
4 False.
5 True.
6 False.
7 True.
8 True.
9 False.

Page 66

1 e
2 b
3 a
4 d
5 h
6 c
7 f
8 g

Page 67 (1)

1 B
2 C
3 A

Page 67 (2)

1 C
2 A
3 B

Page 68

1 (a) 0–15. (b) 65+.
2 Youth dependency.
3 Youth dependency.
4 Aged dependency.
5 Elderly can provide wise advice.

Page 70

Push factors: Few jobs, poor education opportunities, poor health care, drought.
Pull factors: good jobs available, better education, safer environment.
Human barriers to migration: visa requirement, no entitlement to work in destination country.
Physical barriers to migration: Mediterranean Sea, mountains to cross.

Page 71

1 240,000.
2 Syria, Afghanistan, Iraq, Eritrea, Somalia.
3 Idomeni camp and Moria camp.
4 18%.
5 185,000.
6 Limited education for children, dangerous journey paying people-smugglers to take them across the Mediterranean in crowded conditions.
7 Golden Dawn Party.

Page 74

1 (a) CBD.
   (b) CBD.
   (c) Suburbs.
   (d) Inner city.
   (e) Rural–urban fringe.
2 Central business district.
3 Rural–urban fringe.

Page 76

1 Ports, industry, tourism, residential.
2 Shadows would tell you a building is tall.
3 Suburbs.
4 CBD.

Page 78

1 Small, rundown, few facilities, inefficient.
2 Older buildings are bought by developers and redeveloped for people on a high income, such as expensive apartments. This can result in the original residents being excluded.
3 People in the original communities may be pushed out as they cannot afford the high prices for the new apartments. The new jobs in the area, such as programming and service industries, may not be suited to the skills of older original residents, so they may lose their jobs and not be able to find new jobs. Older service industries may not be able to afford the high rents, so original residents may lose their favourite cafes or shops, and may not like the newer style services that replace them.

Page 79

1 Streets are narrow, inner-city areas usually have no driveways so people park on the road, people driving into the CBD may park here to avoid parking charges.
2 23,500.
3 Mexico City.
4 £4 billion.
5 1.4 minutes.

Page 80

1 £160.
2 Los Angeles.
3 Residents-only permit systems.

Page 83

1 Premier Inn, Titanic Studios, Microsoft, Citigroup.
2 7500 apartments and townhouses.
3 All new buildings use energy conservation, solar heating and rainwater harvesting.
4 £385 million.
5 15,000–18,000.

Page 84

1 People, especially migrants, arriving in the city cannot afford to buy houses or pay expensive rent, so they move to a shanty town.
2 Push factor.
3 Pull factor.
4 Bustees.
5 4.5 million.
6 Big, badly built, basic, beside roads, bulldozed, below poverty line.

Page 86

1 GNI per person.
2 Doctors per 1000 population, life expectancy, literacy rate.
3 MEDC life expectancy over 75 years, LEDC under 75 years.
4 MEDC.

Page 87

1 Strength: shows whether a country is rich or poor.
   Weakness: assumes everyone has an equal share in the wealth of the country.
2 Strength: give an idea of the living conditions in the country.
   Weakness: average figures, hide variation in society.
3 Life expectancy, adult literacy/school enrolment, GDP per person.
4 Combines health, education and wealth data to give a rounded picture, shows how a country uses its wealth.

Page 88

1 Historical factors, environmental factors, dependence on primary activities, debt.
2 Historical factors: countries such as the UK and Spain took over countries as colonies. The rich countries took raw materials from the colonies, such as the UK taking cotton from India, and the colony didn't benefit much from it, while the richer country developed further.

Exam practice answers and quick quizzes at **www.hoddereducation.co.uk/myrevisionnotesdownloads**

Environmental factors: many LEDCs suffer hurricanes, earthquakes and floods. These cause destruction and it costs a lot of money to recover, so development in these countries is delayed.

Dependence on primary activities: Zambia gets 98% of its income from exporting copper. This means that if the price of copper goes down, or if it runs out of copper, Zambia will lose most of its income. Also these activities do not make as much money as processing the raw materials, so Zambia has had less money available for development projects

Debt: Ecuador owes $10 billion, so spends a lot of its money on repayments, instead of hospitals and schools which would help it develop.

Page 89

Environmental – 7, 12, 13, 14, 15.

Economic – 1, 2, 8, 9.

Social and cultural – 3, 4, 5, 6, 10, 11, 16, 17.

Page 90 (1)

1 No poverty, decent work and economic growth, life on land – or any others from the list on page 89.

2 No poverty: to give people higher incomes in poor countries, so that the gap between rich and poor is reduced.

Decent work and economic growth: this would increase GDP by 7% a year in the poorest countries, making them grow richer rapidly and reducing the development gap.

Life on land: to manage forests sustainably and protect ecosystems so that countries can benefit from their natural environments. This would bring LEDCs closer to the requirements on MEDCs for conservation.

Page 90 (2)

1 Option (a).

2 Option (a).

3 Option (a).

4 Option (a).

5 Option (b).

6 Option (b).

Page 91

1 90 litres.

2 Ten years.

3 Requires no spare parts, requires no power, lasts up to ten years.

4 More than twenty.

Page 93

1 Globalisation is the way people, goods, money and ideas move around the world faster and more cheaply than ever before.

2 World trade, multinational corporations, economic decision or events in one country affect other countries very quickly.

Page 94

1 Life expectancy has gone up from 59 years in 1990 to 68 in 2015, and adult literacy rates have increased from 50% in 1990 to 74% in 2011.

2 Western-style clothes and behaviour are considered shocking by some. Half of children under five are still malnourished.

Page 96 (1)

1 Travelling by car burns petrol, which comes from oil, which is a fossil fuel made from dead plants and animals, so contains carbon. When it is burned it releases carbon dioxide.

2 Travelling to Spain by plane uses a lot of fuel, which comes from oil, releasing carbon dioxide when it is burned.

3 Grapes have to be transported from Italy, which takes fuel, releasing carbon dioxide when it is burned.

4 Turning the heating up in the house increases the amount of oil or gas used, or electricity, which may be made by burning oil or gas, which releases carbon dioxide when burned.

5 Using your hairdryer uses electricity, which may be made using oil or gas, releasing carbon dioxide.

Page 96 (2)

(a) Caroline Careless.

(b) Gregory Green.

(c) Gregory Green.

(d) Caroline Careless.

(e) Gregory Green.

(f) Gregory Green.

(g) Caroline Careless.

(h) Caroline Careless.

(i) Gregory Green.

(j) Caroline Careless.

Page 97

1 92%.

2 1500 km.

3 Strangford Lough.

4 Melt more quickly because the dark seawater will absorb more of the Sun's rays.

Page 98

1 (a) Disposal.

(b) Recycling.

(c) Reusing.

(d) Reducing.

(e) Composting.

2 Reducing, reusing, composting, recycling, disposal.

Page 99

(a) Reusing.
(b) Reducing.
(c) Recycling.
(d) Reducing.
(e) Recycling.
(f) Reducing.
(g) Recycling.
(h) Reducing/reusing.

Page 103

Cyprus: negative, environmental.
Zanzibar: negative, environmental.
Employment: positive, economic.
Uluru: negative, cultural.
Kenya: negative, economic.

Page 104

Responsible:
- Learning please and thank you in the local language.
- Asked permission for selfies.
- Made sure none of the wrappers from their picnic blew away.
- Ate local food at a local restaurant.

Could have been more responsible:
- Went straight to McDonald's – they could have used a local restaurant.
- Bought elephant tusk necklaces – they could have bought other craft not made from poached elephants.
- Picked flowers – they could have taken photos.
- Had a long shower – they could have had a short shower.

Page 107 (1)

1 (b), (e), (a), (c), (d), (f).
2 Collect data is missing. 'Tom measured the width, depth and bedload size at six points along the river.'

Page 107 (2)

1 Plan fieldwork, Present data, Analysis and interpretation.
2 A hypothesis is a statement you want to test to see if it is correct.
3 Analysis includes patterns, with figures, and anomalies.

Page 108

The aim is a summary of what you are trying to find out, and is quite general. Hypotheses are very specific statements to test to see if they are true. For example, the aim could be to test the Bradshaw model, with a specific hypothesis that the cross-sectional area of the river will increase downstream.

Page 110

1 (a) Primary.
   (b) Primary.
   (c) Secondary.
   (d) Primary.
   (e) Primary or secondary.
   (f) Primary.

Page 114

Answers depend on student's own data.

Page 115

1 Pattern, figures, anomaly, hypothesis.
2 An anomaly.

Page 118

Amount, reliability, clarity.

Index

My Revision Notes CCEA GCSE Geography second edition